NORTH AMERICAN FIELD GUIDES

SONGBIRDS

Golriz Golkar

Field Guides

An Imprint of Abdo Reference | abdobooks.com

CONTENTS

WHAT ARE SONGBIRDS?

As their name suggests, songbirds are known for their strong vocal skills. Their sounds range from harsh calls and screeches, such as those made by crows, to the melodious tunes of thrushes and mockingbirds. Songbirds are found on every continent except Antarctica. Among songbirds of the northern hemisphere, males sing more than females.

There are more than 4,000 species of songbirds, divided into about 50 families. They make up nearly half of all birds. Songbirds come in a variety of sizes, shapes, and colors. They live in a large range of land habitats, from forests to prairies to deserts. Some also live in or near wetlands or stream corridors. Most songbirds are omnivores that eat insects, seeds, berries, and orchard fruits such as figs, apricots, and cherries. Their diet varies with the time of year. Songbirds have a distinctive toe anatomy that lets them perch on tree branches.

WHAT ARE SONGBIRDS LIKE?

Like other birds, songbirds lay eggs. Young songbirds may look different from adults of the same species. Males and females sometimes have different colors and markings. Songbirds molt, or lose feathers and grow new ones. Some species molt more than once a year. When molting, a songbird's colors and markings may look different from others of its species.

Certain songbird families, such as grosbeaks and finches, have specialized adaptations. Their cone-shaped beaks help them crack open seeds. Some North American songbirds migrate within the continent, while others migrate to other parts of the world. Other songbird species spend their whole lives in the same general region.

Songbirds create their music with the help of the syrinx. While most birds have this vocal organ, songbirds have an especially developed one. During an exhale, muscles inside the syrinx are engaged. The muscles control vibrating membranes that create specific sounds when air flows over them. Through physical syrinx differences and greater control of the organ, songbirds create more elaborate sounds than other birds. Some songbirds sing meticulous songs, while others mostly make call sounds.

Songbirds also develop their vocal skills as they grow up. They are born knowing how to sing basic tunes or make calls. They learn more refined sound patterns from older birds. Some songbirds learn only a few new songs or calls when young. Others continue to learn new ones into adulthood.

Songbirds use their calls for many purposes. They use them to defend territory, attract mates, and communicate. Males call during mating season to attract females and warn rival males away from their territory. Some males always sing from a perch, while others also sing in flight. Individual males may sometimes add random notes or change pitch to attract more female attention. Other males may notice and do the same with their own improvised songs. Songbird tunes are therefore always evolving.

HOW TO USE THIS BOOK

Tab shows the songbird category.

The songbird's common name appears here.

CUCKOOS

BLACK-BILLED CUCKOO

(COCCYZUS ERYTHROPTHALMUS)

Black-billed cuckoos have slightly curved bills and long ... have brown backs and heads, white undersides, ... ye-ring. These slow birds sit very still in trees ... They eat fruits, seeds, and insects—with a ... for spiny caterpillars. When spines get stuck in a ... tomach, the cuckoo coughs them up into pellets ... them out. These cuckoos sing a soft, repeated *coo* song.

This paragraph gives information about the songbird.

HOW TO SPOT

Size: 11 to 12.2 inches (28 to 31 cm) long; 1.4 to 2.3 ounces (39.7 to 65.2 g)

Range: Southern Canada, United States, Mexico, and ...al America

...**t:** Forests, woodlands, and thickets

...aterpillars, large insects, fruits, and seeds

Fun Facts give interesting information about the songbird.

FUN FACT

Black-billed cuckoos and yellow-billed cuckoos are sometimes called rain crows. According to old folktales, they often call just before rainfall.

WHAT ARE CUCKOOS?

Of 147 cuckoo species in the world, only three live in North America. Cuckoos have long tails, stout and downcurved bills, and outer toes that point backward. They are brood parasites. Yellow-billed and black-billed cuckoos migrate to South America in the fall. Mangrove cuckoos stay close to home in Florida.

Sidebars provide additional information about the topic.

YELLOW-BILLED CUCKOO

(COCCYZUS AMERICANUS)

The songbird's scientific name appears here.

Yellow-billed cuckoos have thick yellow bills and long tails. Their bills are almost as long as their heads. They have brown heads, black on the undersides of their tails, and a yellow eye-ring. These birds live in wooded habitats such as farms and thickets. They often live in areas with streams or marshes nearby, feasting on cicadas, katydids, and many fruits and seeds. Yellow-billed cuckoos make knocking and cooing calls at any time of day.

Images show the songbird.

HOW TO SPOT

Size: 10.2 to 11.8 inches (26 to 30 cm) long; 1.9 to 2.3 ounces (53.9 to 65.2 g)

Range: Southern Canada, United States, Mexico, Central America, and Cuba

Habitat: Woodlands, orchards, and thickets

Diet: Caterpillars, other insects, fruits, and seeds

45

How to Spot features give information about the songbird's size, range, habitat, and diet.

BELL'S VIREO *(VIREO BELLII)*

Bell's vireos are found throughout North America. Their bills are smaller than those of other vireos, and they are often used to pluck insects from vegetation. Western birds have grayish heads and backs, white undersides, and pale wing bars. Eastern birds have greenish backs and yellow sides with more defined wing bars. Many of these birds are losing their habitat because of wildfires, agriculture, and grazing. Males are not always easy to spot, but their loud voices are often heard. They sing hurried and scratchy songs.

HOW TO SPOT

Size: 4.5 to 4.9 inches (11.4 to 12.4 cm) long; 0.26 to 0.35 ounces (7.4 to 10 g)

Range: Southern Canada, United States, and Central America

Habitat: Shrublands, chaparral, grasslands, and scrub oaks

Diet: Insects, spiders, and small berries

Chaparral

Western Bell's vireo

WHAT ARE VIREOS?

Vireos are small, round birds with hooked bills. All 50 species are found in the western hemisphere. Vireos build cup-like nests from leaves, bark, and spider silk. Most vireos migrate south in the fall, flying as far as South America. Migratory vireos fly back north in spring.

BLACK-CAPPED VIREO

(VIREO ATRICAPILLA)

Black-capped vireos have thick bills, short wings, and medium-length tails. Males have shiny black heads, white throats, and white circles around bright red eyes. They have olive-green backs, white undersides with yellow markings, and white wing bars. Females are much paler in color. These rare birds are found mostly in Texas, migrating to Mexico in the fall. Black-capped vireos stay low to the ground, searching for insects and plants to eat. These birds sometimes hang upside down from branches to check leaves for insect larvae.

FUN FACT

Black-capped vireos sing long songs that go up and down in tone. They use about ten times more phrases than other vireos.

HOW TO SPOT

Size: 4.3 inches (11 cm) long; 0.3 ounces (8.5 g)

Range: Southern United States and Mexico

Habitat: Canyons, forests, and scrub oaks

Diet: Butterflies, beetles, spiders, larvae, and plants

Female

Insect larvae

BLUE-HEADED VIREO

(VIREO SOLITARIUS)

Blue-headed vireos have medium-length tails and thick legs. Their heads and backs are streaked with blue, gray, green, and yellow. They have white eye-rings. Their undersides are white, and their wings are black and white. Vireos make their homes in the scrubby vegetation of coniferous, deciduous, or mixed forests. They hop from tree to tree when looking for prey. They also enjoy fruits of elder, sumac, and dogwood trees. Their songs are high-pitched and slow.

FUN FACT

Male blue-headed vireos select a nesting site. If a female likes it, the male starts building a nest—but the female finishes the job.

HOW TO SPOT

Size: 4.7 to 5.9 inches (12 to 15 cm) long; 0.5 to 0.6 ounces (14.2 to 17 g)

Range: Canada, United States, and Central America

Habitat: Coniferous and deciduous forests

Diet: Insects, spiders, snails, and fruits

CASSIN'S VIREO *(VIREO CASSINII)*

Cassin's vireos have short and thick bills. This bird has an olive-green body and a gray head with white eye-rings. Undersides and wing bars are white. Cassin's vireos search for prey in trees. Sometimes they fly after prey to capture them. They eat a variety of insects, including butterflies and caterpillars. Fruits and seeds make up their winter diet. They live in many types of forests, from deciduous oak forests near coasts to coniferous or mixed forests in inland areas. Their songs are full of short pauses and make a whirring sound.

Deciduous forest

HOW TO SPOT

Size: 4.3 to 5.3 inches (11 to 13.5 cm) long; 0.5 to 0.6 ounces (14.2 to 17 g)

Range: Western Canada, northwestern United States, and Mexico

Habitat: Coniferous forests, deciduous forests, and mixed forests

Diet: Insects, spiders, small fruits, and seeds

Coniferous forest

GRAY VIREO *(VIREO VICINIOR)*

Gray vireos have short wings with faint wing bars and eye-rings. Their heads and backs are gray, and their undersides are white. They blend in well with their surroundings. Gray vireos look for insects in dense brush close to the ground. The insects they eat can be very large, such as grasshoppers, cicadas, and caterpillars. Gray vireos overtake these prey by pounding them against branches. They may remove the head or wings before eating them. This bird sings long whistling songs.

HOW TO SPOT

Size: 5.1 to 5.9 inches (13 to 15 cm) long; 0.4 to 0.5 ounces (11.3 to 14.2 g)

Range: Southwestern United States and northwestern Mexico

Habitat: Desert scrublands, chaparral, and canyons

Diet: Insects and small fruits

HUTTON'S VIREO *(VIREO HUTTONI)*

Hutton's vireos have big heads and long tails. They have olive-green-and-gray heads, backs, and undersides. Their dark gray wings have two white wing bars. Those that live near the coast have richer colors than inland birds. Hutton's vireos mostly live in evergreen oak, coniferous, and mixed forests. Unlike most vireos, the Hutton's vireo does not migrate. They search slowly for insects and other food in trees. During nonbreeding times, they fly with large mixed flocks of woodland songbirds.

FUN FACT

The Hutton's vireo sings a repeated song that sounds like a rising *zu-wee* or a falling *zo-zoo.*

HOW TO SPOT

Size: 3.9 to 4.7 inches (10 to 12 cm) long; 0.3 to 0.5 ounces (8.5 to 14.2 g)

Range: Southwestern Canada, western United States, northwestern Mexico, and Central America

Habitat: Oak forests, coniferous forests, mixed forests, and chaparral

Diet: Insects, spiders, and plants

RED-EYED VIREO *(VIREO OLIVACEUS)*

Red-eyed vireos have angular heads and long bills. Their heads, backs, and undersides are olive green and yellow. Adults have red eyes, while younger birds have dark eyes. All red-eyed vireos have a white stripe above the eyes. It has the largest habitat range of all vireos. Their diet changes throughout the year. Insects make up most of their spring and summer diet, and fruits are added in the fall and winter. Red-eyed vireos often sing continuously in the summer. They make repetitive musical phrases in a question-and-answer pattern.

HOW TO SPOT

Size: 4.7 to 5.1 inches (12 to 13 cm) long; 0.4 to 0.9 ounces (11.3 to 25.5 g)

Range: Southern Canada, United States, eastern Mexico, and Central America

Habitat: Deciduous forests, woodlands, and rain forests

Diet: Insects, spiders, snails, seeds, and fruits

WARBLING VIREO *(VIREO GILVUS)*

Warbling vireos have olive-green-and-gray heads and backs and a white line over the eyes. Their undersides are white, and they have yellow-and-brown wings. They live in mostly deciduous forests near bodies of water. These vireos also nest in residential areas and city parks. The warbling vireo hunts for insect prey in treetops, often smacking them for capture. They add fruits to their diet in fall and winter. These highly musical birds sing a fast, loud, and cheerful song.

HOW TO SPOT

Size: 4.7 to 5.1 inches (12 to 13 cm) long; 0.3 to 0.6 ounces (8.5 to 17 g)

Range: Canada, United States, Mexico, and Central America

Habitat: Deciduous forests, urban parks, and woodlands

Diet: Caterpillars, moths, butterflies, small insects, spiders, and berries

WHITE-EYED VIREO *(VIREO GRISEUS)*

White-eyed vireos have gray heads. Yellow eye-rings circle their white eyes with black pupils. The chest, throat, and wing bars are white. Their undersides are yellow. The dark eyes of young birds turn white during their first winter or spring. These birds swallow small prey such as flies and moths. Larger prey such as butterflies and spiders are pinned down with the feet before being eaten. White-eyed vireos sometimes sing all day. Both females and males sing in wintering ranges, but only males sing in breeding ranges.

FUN FACT

White-eyed vireos wash themselves in the morning by rubbing their bodies against dewy leaves and bushes.

HOW TO SPOT

Size: 4.3 to 5.1 inches (11 to 13 cm) long; 0.3 to 0.5 ounces (8.5 to 14.2 g)

Range: Southeastern United States, Central America, and Bahamas

Habitat: Brushlands, forest edges, thickets, and mangroves

Diet: Large insects and spiders

YELLOW-THROATED VIREO

(VIREO FLAVIFRONS)

Yellow-throated vireos have big heads and short tails. They have olive-green heads and backs, white undersides, and yellow eye-rings. The throat and chest are yellow. These vireos frequently change habitats. They live in deciduous forests during breeding season. Woodlands and shrubs are preferred during migration, while rain forests provide a winter home. These vireos hop between tree branches, picking insects, fruits, and seeds from twigs and leaves. Yellow-throated vireos sing short songs with repeated sounds.

HOW TO SPOT

Size: 5.1 to 5.9 inches (13 to 15 cm) long; 0.5 to 0.7 ounces (14.2 to 19.8 g)

Range: Southern Canada, United States, Mexico, Central America, and Cuba

Habitat: Deciduous forests, woodlands, and rain forests

Diet: Insects, spiders, fruits, and seeds

FUN FACT

Unlike many male North American songbirds, male yellow-throated vireos help keep the eggs warm and raise their young.

BAY-BREASTED WARBLER

(SETOPHAGA CASTANEA)

Bay-breasted warblers are small and chubby. These warblers have tiny pointed bills, long wings, and two white wing bars. In the summer, they have red-and-brown heads, gray backs, and beige undersides. In the fall, the back and head become green and yellow while the undersides turn white. They breed in spruce, fir, and other coniferous forests. Bay-breasted warblers feast on spruce budworms. They choose any forest where spruce budworms are available. They sing a high, single-pitched song that is hard to hear from far away.

HOW TO SPOT

Size: 5.5 inches (14 cm) long; 0.3 to 0.6 ounces (8.5 to 17 g)
Range: Canada, United States, Mexico, and Central America
Habitat: Woodlands, boreal forests, and forest edges
Diet: Insects and berries

FUN FACT

Bay-breasted warblers usually search for spruce budworms at low tree levels. This way, they have less competition with other birds looking for these worms.

WHAT ARE WARBLERS?

Warblers are found on every continent except Antarctica. The more than 50 North American species are often called wood warblers. These warblers are smaller and more colorful than those in other regions. In winter, they migrate south to tropical regions. They migrate back north in spring.

BLACK-AND-WHITE WARBLER

(MNIOTILTA VARIA)

Black-and-white warblers have long bills and wings, a short tail, and black-and-white stripes. They have heavier legs and a longer hind claw than other wood warblers. These features allow them to cling to tree bark as they look for insect prey. They crawl along deciduous and mixed-forest trees to look for prey. In winter, they live in many habitats including wetlands and other forest types. The black-and-white warbler defends its territory. These birds can be aggressive when protecting their nests and food, even attacking other birds of their kind. They sing squeaky songs.

HOW TO SPOT

Size: 4.3 to 5.1 inches (11 to 13 cm) long; 0.3 to 0.5 ounces (8.5 to 14.2 g)

Range: Canada, United States, Mexico, Central America, and Cuba

Habitat: Deciduous forests and mixed forests

Diet: Insects, ants, and larvae

BLACK-THROATED BLUE WARBLER *(SETOPHAGA CAERULESCENS)*

Black-throated blue warblers are small, chubby birds with pointed bills. Males have midnight-blue backs and heads, black throats and faces, and white undersides. Females have gray-and-olive bodies. Both sexes have a small white patch on their wings. These warblers live in deciduous forests where they search for prey under shrubs. During migration, they visit woodlands and gardens. Winters are spent in rain forests and woodlands. This warbler's songs vary from slow and buzzy to high-pitched. Males sing to defend their territories and chase away rival males.

Female

Male

FUN FACT

Males and females look so different from each other that they were once described as two different bird species.

HOW TO SPOT

Size: 4.3 to 5.1 inches (11 to 13 cm) long; 0.3 to 0.4 ounces (8.5 to 11.3 g)

Range: Southern Canada, United States, Mexico, Central America, and Caribbean

Habitat: Deciduous forests, woodlands, and rain forests

Diet: Spiders, flies, and caterpillars

BLACK-THROATED GRAY WARBLER *(SETOPHAGA NIGRESCENS)*

Black-throated gray warblers are small and chubby with round bills. They have large heads and a yellow spot above the eye. Streaks of gray, black, and white cover the body. Females are slightly paler. These warblers move more slowly than other warbler species. They take their time looking for food, hopping from branch to branch in search of prey. They live in coniferous forests such as pine and pine-oak. Their song is consistent and recognizable, sounding like *zeedle zeedle zeedle zeet-chee*.

Male

HOW TO SPOT

Size: 4.3 to 5.1 inches (11 to 13 cm) long; 0.25 to 0.35 ounces (7 to 10 g)

Range: Western Canada, United States, and Mexico

Habitat: Coniferous forests, woodlands, and thickets

Diet: Insects

BROOD PARASITES

Brood parasites are birds that lay eggs in the nests of other bird species. Cowbirds and cuckoos are two examples found in North America. They expect other birds to raise their young. Some species watch over the eggs and raise the nestlings, while others destroy them or send the nestlings away.

CAPE MAY WARBLER

(SETOPHAGA TIGRINA)

Cape May warblers are round, short-tailed birds. Adult males have olive-green-and-yellow bodies with rufous cheeks. Females and young males are duller with a yellow-green rump. These warblers live in boreal forests such as balsam fir and spruce during breeding season. They enjoy many forest types and shrubby habitats during migration and winters. Cape May warblers use their curled tongues to drink nectar and pluck insects off flowers. Cape May warblers sing a series of high-pitched notes at the same pitch and volume.

Breeding male

FUN FACT

Cape May warblers have a unique tongue among warblers. It is curled and half-tube shaped, helping the bird sip nectar.

Female

HOW TO SPOT

Size: 4.7 to 5.1 inches (12 to 13 cm) long; 0.4 to 0.5 ounces (11.3 to 14.2 g)

Range: United States, Canada, Caribbean, and Central America

Habitat: Boreal forests and shrublands

Diet: Spruce budworms, spiders, insects, and flower nectar

CHESTNUT-SIDED WARBLER

(SETOPHAGA PENSYLVANICA)

Chestnut-sided warblers are slim warblers with a long tail. Young birds and nonbreeding adults have bright green upper bodies, white-gray bellies, and two wing bars. A white eye-ring circles the eyes. Breeding adults have gray to white bodies with a yellow head. The face has black marks, and the sides are reddish-brown, or chestnut-colored. These warblers hop along branches looking for insects under leaves. Males sing two different songs. One is a short, accented, and high-pitched song for attracting females. The other is calmer and is sung during nesting.

Breeding male

Breeding female

HOW TO SPOT

Size: 4.7 to 5.5 inches (12 to 14 cm) long; 0.4 to 0.5 ounces (11.3 to 14.2 g)

Range: Southeastern Canada, midwestern and eastern United States, Mexico, Central America

Habitat: Deciduous forests and thickets

Diet: Insects

HERMIT WARBLER

(SETOPHAGA OCCIDENTALIS)

Hermit warblers are small and chubby with a long tail. Their bills are short and straight. Breeding males have a bright yellow head, a black throat, a dark gray back, and two white wing bars. Females and nonbreeding males have duller colors. Hermit warblers live in coniferous forests in spring and summer. These warblers search for insects high up in trees but also eat fruits in winter. They also drink honeydew, a sweet secretion made by insects and aphids. Their songs have a mix of high-pitched and buzzy sounds.

HOW TO SPOT

Size: 5.5 inches (14 cm) long; 0.3 to 0.5 ounces (8.5 to 14.2 g)

Range: Western and northeastern United States, Mexico, and Central America

Habitat: Coniferous forests, woodlands, and desert oases

Diet: Insects, spiders, and fruits

Breeding male

WILSON'S WARBLER

(CARDELLINA PUSILLA)

Wilson's warblers are among the smallest warblers. They have long, thin tails, and their bills are small and thin. Their heads and backs are yellow and olive green, and their undersides are yellow. Males have a black cap on their heads. These warblers breed in forests and mountains. They pick off insects from tree leaves and twigs. They also drink honeydew from oak trees during winter. Wilson's warblers are very active, always hopping, dancing, and singing sweet and simple songs.

Male

FUN FACT

Wilson's warblers pass through all of the US states except Alaska and Hawaii during fall and spring migration.

HOW TO SPOT

Size: 3.9 to 4.7 inches (10 to 12 cm) long; 0.2 to 0.3 ounces (5.7 to 8.5 g)

Range: United States, Canada, Mexico, and Central America

Habitat: Thickets, coniferous forests, and forest edges

Diet: Spiders, flies, bees, and other insects

WORM-EATING WARBLER

(HELMITHEROS VERMIVORUM)

Worm-eating warblers are small birds with long and spiky tails, long wings, and sharp bills. Their heads, backs, and undersides are brown and green. Black lines mark their heads. Worm-eating warblers live in deciduous forests and woodlands for breeding. They also live in mangroves, rain forests, and mountain forests during migration and winters. These warblers eat caterpillars and slugs, but they don't actually eat earthworms. Their song is a trill that has a dry sound and has been described as insect-like.

HOW TO SPOT

Size: 4.4 to 5.2 inches (11.2 to 13.2 cm) long; 0.4 to 0.5 ounces (11.3 to 14.2 g)

Range: United States, Mexico, Central America, and Cuba

Habitat: Deciduous forests, mixed forests, and mangroves

Diet: Caterpillars, grubs, slugs, and spiders

YELLOW WARBLER

(SETOPHAGA PETECHIA)

Yellow warblers are common in North America. These small birds have round heads and straight, thin bills. They have yellow bodies and black eyes. Males are a deeper yellow than females. These warblers spend breeding season in thickets near streams and wetlands. Winters are spent in marshes and forests. Yellow warblers hop constantly from branch to branch, searching for insects. They sing a sweet whistling song that is often heard in spring and summer.

Male

Mangroves

HOW TO SPOT

Size: 4.7 to 5.1 inches (12 to 13 cm) long; 0.3 to 0.4 ounces (8.5 to 11.3 g)

Range: Canada, United States, Mexico, Central America, and Cuba

Habitat: Woodlands, mangroves, marshes, and forests

Diet: Caterpillars, beetles, and wasps

BEWICK'S WREN

(THRYOMANES BEWICKII)

Bewick's wrens have slender bodies. Their heads are pale brown and gray with a white stripe over the eye. Their backs and wings are brown, and their undersides are white. Bewick's wrens live in many woodland and scrubland habitats. They eat insects and their larvae. In winter, fruits, seeds, and nuts are added to their diet. An adult Bewick's wren will sometimes eat mud and pebbles to help with digestion. The Bewick's wren makes clear high notes and musical trills.

HOW TO SPOT

Size: 5.3 inches (13.5 cm) long; 0.3 to 0.4 ounces (8.5 to 11.3 g)

Range: Southwestern Canada, United States, and Mexico

Habitat: Woodlands, desert scrublands, and thickets

Diet: Insects, spiders, fruits, and seeds

WHAT ARE WRENS?

Wrens are small and chunky birds. The 85 wren species are located in North, Central, and South America. They have slightly downcurved bills and short, rounded wings. Their tails are short and upright. Some wrens fly south in the fall, returning home in spring.

CACTUS WREN

(CAMPYLORHYNCHUS BRUNNEICAPILLUS)

Cactus wrens are large wrens with long tails. They have brown-and-white bodies, a black-and-white streaked tail, and a speckled chest. These wrens are very active—if they are not searching for food, they are singing, fanning their tails, or arguing with other birds. Cactus wrens build their nests on cacti spikes to keep predators away. During cool mornings, they search for prey on the ground. When it's hot, they search for shade under shrubs and cacti. As desert dwellers, they can survive without drinking water.

FUN FACT

The cactus wren makes a funny call sound. It sounds like an old car with an engine that won't start!

HOW TO SPOT

Size: 8 to 9 inches (20 to 23 cm) long; 1.3 to 1.5 ounces (36.9 to 42.5 g)

Range: Southwestern United States and Mexico

Habitat: Desert scrublands and shrublands

Diet: Insects, fruits, and seeds

CANYON WREN

(CATHERPES MEXICANUS)

Canyon wrens are potbellied wrens. They have bodies that are gray and brown with white speckles. Their tails are rufous, and their wings and tails have black streaks. When searching for prey, they sometimes climb vertical cliffs or use their bills to probe for insects. Like the cactus wren, these wrens get their water from food. Canyon wrens sing a loud and sweet whistling tune that echoes in the canyons where they live. Males sing from rocky perches in spring and summer, and females sometimes join them.

HOW TO SPOT

Size: 4.5 to 6.1 inches (11.4 to 15.5 cm) long; 0.3 to 0.7 ounces (8.5 to 19.8 g)

Range: Southwestern Canada, western and midwestern United States, and Mexico

Habitat: Canyons

Diet: Insects and spiders

FUN FACT

Canyon wrens sometimes grab insects from wasp nests or spiderwebs.

CAROLINA WREN

(THRYOTHORUS LUDOVICIANUS)

Carolina wrens have long tails. They have rufous heads and backs and pale orange undersides. They have a white stripe above the eye. These wrens look for food by poking around shrubs and climbing tree trunks. They sometimes eat larger prey such as small snakes and frogs. These musical birds sing constantly to defend their territory. If an intruder approaches, a Carolina wren makes lots of noise. These birds stay mostly in the same place year-round and do not migrate.

HOW TO SPOT

Size: 4.7 to 5.5 inches (12 to 14 cm) long; 0.6 to 0.8 ounces (17 to 22.7 g)

Range: United States, Central America, and Mexico

Habitat: Woodlands, thickets, and swamps

Diet: Insects, spiders, lizards, frogs, and snakes

HOUSE WREN

(TROGLODYTES AEDON)

House wrens are common across most of the western hemisphere. They have flat heads and slightly long tails. Their bodies are pale brown, although their wings and tails are darker. During breeding season, they sing high-pitched, cheerful songs from visible spots such as bush edges. In winter, they hide in hedges and thickets. These wrens build twig nests in cavities found in backyards, farms, and buildings. They also use human-made nests made from similar materials.

HOW TO SPOT

Size: 4.3 to 5.1 inches (11 to 13 cm) long; 0.3 to 0.4 ounces (8.5 to 11.3 g)

Range: Canada, United States, Mexico, and Central America

Habitat: Deciduous forests, coniferous forests, yards, farms, and swamps

Diet: Insects, spiders, and snail shells

PROTECTING THE YOUNG

Parasites such as mites are a threat to house wren nestlings since they may eat these young birds. For this reason, house wrens often add spider egg sacs to their nests. When the baby spiders hatch, they eat any parasites in the nest—keeping the baby birds safe.

MARSH WREN

(CISTOTHORUS PALUSTRIS)

Marsh wrens have rufous bodies streaked with black and white. They have a white stripe above the eye. Marsh wrens pick spiders and insects from leaves and plants, especially in cattail marshes. They mostly stay hidden in their habitat, but they make plenty of noise. They are fierce birds that pierce the eggs of other birds and fight for food. Marsh wrens sing in the morning and the evening, and sometimes they sing all night. They make a fast, gurgling, and shrill sound. Western marsh wrens are paler and a little less musical than eastern ones.

HOW TO SPOT

Size: 3.9 to 5.5 inches (10 to 14 cm) long; 0.3 to 0.5 ounces (8.5 to 14.2 g)

Range: Southern Canada, United States, and Mexico

Habitat: Marshes and thickets

Diet: Insects and spiders

PACIFIC WREN

(TROGLODYTES PACIFICUS)

Pacific wrens are among the smallest wrens in North America. They have brown bodies with dark markings on their wings, tails, and undersides. These wrens look for food and build nests near water, under shrubs, or near logs and dead trees. They pick insects off bark, but they also eat juniper berries during nonbreeding season. Pacific wrens are lively birds. They bounce their heads often and beat their wings fast in flight. They sing a loud, high-pitched, whistling song.

FUN FACT

Pacific wrens sometimes gather near streams in the Pacific Northwest when salmon are migrating there. Salmon carcasses attract insects that the wrens eat.

HOW TO SPOT

Size: 3.1 to 4.7 inches (8 to 12 cm) long; 0.3 to 0.4 ounces (8.5 to 11.3 g)

Range: Western Canada, Pacific Northwest United States, and western United States

Habitat: Evergreen forests, deciduous forests, urban parks, and gardens

Diet: Spiders, insects, and berries

ROCK WREN

(SALPINCTES OBSOLETUS)

Rock wrens have short, speckled wings and long tails. They have brown bodies with gray specks on their backs. These wrens live in a wide variety of dry and rocky habitats, from cliffs to canyon walls and even construction sites. They use their long bills to grab insect prey from the ground and rock crevices. Like some other wrens, they do not drink water. Rock wrens make loud and prickly sounds with repeated phrases.

FUN FACT

Rock wrens build nests inside hidden rock crevices. They also build paths leading to their nests. The paths can be made of pebbles, bones, and human-made objects.

HOW TO SPOT

Size: 4.9 to 5.9 inches (12.4 to 15 cm) long; 0.5 to 0.6 ounces (14.2 to 17 g)

Range: Southern Canada, United States, Mexico, and Central America

Habitat: Deserts, valleys, mountains, and grasslands

Diet: Insects, spiders, plants, and seeds

SEDGE WREN

(CISTOTHORUS STELLARIS)

Sedge wrens are very small wrens with long legs and medium-length tails. Their beige bodies are streaked with brown, black, red, and gray. These wrens live in wet fields in grasslands and shallow marshes where they look for insects and spiders. Their specially adapted bills help them pick prey out of dense plants. Sedge wrens are one of the most nomadic birds in North America, constantly changing location. This wren's song is simple compared with those of other wrens: a few chirps followed by a trill.

HOW TO SPOT

Size: 3.9 to 4.7 inches (10 to 12 cm) long; 0.25 to 0.35 ounces (7 to 10 g)

Range: Canada, midwestern and eastern United States, and northern Mexico

Habitat: Marshes and grasslands

Diet: Insects and spiders

WINTER WREN

(TROGLODYTES HIEMALIS)

Winter wrens are light brown with dark brown streaks on the wings, undersides, and tail. They are nearly identical to the Pacific and Eurasian wrens. These active birds hop around stream areas looking for insects under logs, tree roots, and bark. In winter, they fly south for warmer temperatures—often choosing gardens and shrubby habitats as well as forests. During spring, males perch in trees and sing long, high-pitched trills.

HOW TO SPOT

Size: 3.1 to 4.7 inches (8 to 12 cm) long; 0.3 to 0.4 ounces (8.5 to 11.3 g)

Range: Canada and United States

Habitat: Coniferous forests, deciduous forests, and evergreen forests

Diet: Insects, spiders, and fruits

FUN FACT

Based on each bird's weight, the winter wren's sound is ten times more powerful than a rooster's crow. Their bodies shake when they sing!

AMERICAN CROW

(CORVUS BRACHYRHYNCHOS)

American crows are black all over. When they molt, old brownish feathers become shiny new black ones. Their diet includes fruits, insects, and carrion. They are also nest predators that eat the young of many bird species. These social birds travel in large flocks but become aggressive when fighting for food. Excellent problem-solvers, they will turn over objects and pick through trash to find a meal. They live in natural open spaces and human-created areas. American crows are known for their hoarse cawing sound.

HOW TO SPOT

Size: 15.8 to 20.9 inches (40 to 53 cm) long; 11.2 to 21.9 ounces (317.5 to 620.9 g)

Range: Canada and United States

Habitat: Woodlands, farms, urban parks, and marshes

Diet: Earthworms, mice, fruits, grains, fish, and insects

WHAT ARE CORVIDS?

Corvids make up more than 120 species found all over the world. They include jays, crows, and magpies. Corvids have widely different sizes and features. They are typically large with stout bills and plain feathers. These smart, social birds make loud and harsh sounds. Most corvids do not migrate.

BLACK-BILLED MAGPIE

(PICA HUDSONIA)

Black-billed magpies are medium-sized birds. They have long, diamond-shaped tails and strong black bills. Their bodies are black with blue-and-white streaks and patches. These social birds often perch on road signs, fences, and treetops near residential areas. They eat fruits, grains, and insects. Sometimes they kill voles and squirrels or grab baby birds from nests. They also steal food from foxes and coyotes. Black-billed magpies glide long distances in the air and walk with a strut. They make loud and scratchy sounds.

HOW TO SPOT

Size: 17.7 to 23.6 inches (45 to 60 cm) long; 5.1 to 7.4 ounces (144.6 to 209.8 g)

Range: Canada and northwestern United States

Habitat: Grasslands and sagebrush

Diet: Insects, fruits, grains, small mammals, and carrion

BLUE JAY *(CYANOCITTA CRISTATA)*

Blue jays are big-crested birds with wide, rounded tails. Their heads and backs have shades of blue, black, and white. They mostly feast on insects, nuts, and seeds that they hold with their feet and peck open. On occasion, they will eat bird eggs or carrion. They are very common songbirds in North America, often found in forests and backyards. Many blue jays stay in the same range year-round. Others migrate south in the fall but do not leave North America. Blue jays make many different call sounds that travel far distances.

HOW TO SPOT

Size: 9.8 to 11.8 inches (25 to 30 cm) long; 2.5 to 3.5 ounces (70.9 to 99.2 g)

Range: Eastern and central United States and southern Canada

Habitat: Forests, forest edges, and yards

Diet: Insects, acorns, grains, and fruits

FUN FACT

Blue jays are very smart. They mimic the sound of hawks, possibly warning other blue jays if a real hawk is nearby.

COMMON RAVEN *(CORVUS CORAX)*

Common ravens have completely black bodies, including their eyes, legs, and bills. These large birds are found throughout the northern hemisphere and live in many habitats. Ravens are usually found alone or in pairs. These intelligent birds work together to attack prey, with one bird distracting an animal while another attacks. They eat many kinds of food, from mice to herons to human and pet food. These ravens make a deep, gurgling, croaking sound.

FUN FACT
Ravens have unique flying patterns. They can fly upside down and often somersault in the air.

HOW TO SPOT

Size: 22.1 to 27.2 inches (56 to 69 cm) long; 24.3 to 57.3 ounces (689 to 1624.4 g)
Range: Canada, United States, and Mexico
Habitat: Deserts, chaparral, grasslands, and forests
Diet: Small animals, carrion, insects, fish, grains, and berries

PINYON JAY

(GYMNORHINUS CYANOCEPHALUS)

Pinyon jays are medium-sized birds with dagger-like bills, short tails, and no crest. They have dusky blue backs and heads with pale blue-and-gray undersides. These birds travel and hunt in large noisy flocks, looking for food in trees and on the ground. Pinyon jays are more nomadic than migratory, often searching for habitats with plenty of pinyon pine seeds. They make crow-like nasal sounds to communicate with other pinyon jays.

FUN FACT

Pinyon jays have expanding throats. This helps them hold up to 40 seeds, which they store to eat later.

HOW TO SPOT

Size: 10.2 to 11.4 inches (26 to 29 cm) long; 3.2 to 4.2 ounces (90.7 to 119 g)

Range: Northwestern and western United States

Habitat: Woodlands and scrub oaks

Diet: Pinyon pine seeds, fruits, and insects

YELLOW-BILLED MAGPIE

(PICA NUTTALLI)

Yellow-billed magpies have big heads, long tails, curved yellow bills, and rounded wings. The head and back are black, and the undersides are white. These social birds live in the same location year-round. They nest in colonies in tall tree groves. These birds eat mostly acorns in fall and winter. They also eat fruit and insects in spring and summer. Since their range is limited, these birds have been affected by diseases and habitat loss. They are slowly becoming endangered. These birds do not sing but make babbling call sounds.

HOW TO SPOT

Size: 16.9 to 21.3 inches (43 to 54 cm) long; 5.3 to 6 ounces (150.3 to 170 g)

Range: California

Habitat: Grasslands, pastures, orchards, and savannas

Diet: Nuts, fruits, insects, and carrion

Pasture

BLACK-BILLED CUCKOO

(COCCYZUS ERYTHROPTHALMUS)

Black-billed cuckoos have slightly curved bills and long tails. They have brown backs and heads, white undersides, and a red eye-ring. These slow birds sit very still in trees and scrubs. They eat fruits, seeds, and insects—with a preference for spiny caterpillars. When spines get stuck in a cuckoo's stomach, the cuckoo coughs them up into pellets and spits them out. These cuckoos sing a soft, repeated *coo* song.

HOW TO SPOT

Size: 11 to 12.2 inches (28 to 31 cm) long; 1.4 to 2.3 ounces (39.7 to 65.2 g)

Range: Southern Canada, United States, Mexico, and Central America

Habitat: Forests, woodlands, and thickets

Diet: Caterpillars, large insects, fruits, and seeds

FUN FACT

Black-billed cuckoos and yellow-billed cuckoos are sometimes called rain crows. According to old folktales, they often call just before rainfall.

WHAT ARE CUCKOOS?

Of 147 cuckoo species in the world, only three live in North America. Cuckoos have long tails, stout and downcurved bills, and outer toes that point backward. They are brood parasites. Yellow-billed and black-billed cuckoos migrate to South America in the fall. Mangrove cuckoos stay close to home in Florida.

YELLOW-BILLED CUCKOO

(COCCYZUS AMERICANUS)

Yellow-billed cuckoos have thick yellow bills and long tails. Their bills are almost as long as their heads. They have brown heads, black on the undersides of their tails, and a yellow eye-ring. These birds live in wooded habitats such as farms and thickets. They often live in areas with streams or marshes nearby, feasting on cicadas, katydids, and many fruits and seeds. Yellow-billed cuckoos make knocking and cooing calls at any time of day.

HOW TO SPOT

Size: 10.2 to 11.8 inches (26 to 30 cm) long; 1.9 to 2.3 ounces (53.9 to 65.2 g)

Range: Southern Canada, United States, Mexico, Central America, and Cuba

Habitat: Woodlands, orchards, and thickets

Diet: Caterpillars, other insects, fruits, and seeds

AMERICAN TREE SPARROW

(SPIZELLOIDES ARBOREA)

American tree sparrows have round heads and long, skinny tails. They have light gray-and-brown bodies with rusty heads and wings. American tree sparrows nest and breed on the ground in thickets. These sparrows hunt for food in flocks. Even during blizzards, they search hard for seeds hidden in snow. They eat plants and fruits from fall through spring, adding insects to their diet during summer migration. These musical birds sing sweet, high-pitched whistling songs.

FUN FACT

The American tree sparrow must eat 30 percent of its body weight and drink nearly as much per day to survive.

HOW TO SPOT

Size: 5.5 inches (14 cm) long; 0.5 to 1 ounces (14.2 to 28.3 g)
Range: Canada and United States
Habitat: Forests, marshes, thickets, and tundra
Diet: Insects, fruits, seeds, and plants

WHAT ARE SPARROWS?

Sparrows are small, mostly gray-and-brown birds. Their short, thick cone-shaped bills are perfect for cracking seeds. There are more than 30 species in North America. Sparrows in colder regions may migrate south in the fall, searching for warmth.

BACHMAN'S SPARROW

(PEUCAEA AESTIVALIS)

Bachman's sparrows have big, round bills and long, rounded tails. They have brown-and-gray bodies streaked with red. They walk and hop on the ground looking for grass seeds and insects. They often hide in shrubs when hot and remain hidden for long periods. These birds live in burned forest patches but move away if there is too much shrubbery. Their population is decreasing because of logging. This sparrow's song is broken into two parts—a whistle followed by a musical trill.

FUN FACT

To escape predators such as hawks or snakes, these sparrows sometimes hide in underground burrows dug by other animals.

HOW TO SPOT

Size: 4.9 to 6 inches (12.4 to 15.2 cm) long; 0.6 to 0.8 ounces (17 to 22.7 g)

Range: Southeastern United States

Habitat: Woodlands and scrub oaks

Diet: Grass seeds and insects

BAIRD'S SPARROW

(CENTRONYX BAIRDII)

Baird's sparrows have flat heads, big bills, and short tails. Their light-brown bodies are speckled with black and dark brown. They have a touch of pale yellow on their heads. These sparrows spend most of their time on the ground nesting and searching for food. They are nomadic birds that change breeding sites every year. While Baird's sparrows were once common, their population is declining due to habitat loss from agriculture. They sing a clear tinkling song.

HOW TO SPOT

Size: 4.7 to 5.5 inches (12 to 14 cm) long; 0.5 to 0.8 ounces (14.2 to 22.7 g)

Range: Southern Canada, northern and southwestern United States, and Mexico

Habitat: Grasslands

Diet: Insects, spiders, and seeds

BELL'S SPARROW

(ARTEMISIOSPIZA BELLI)

Bell's sparrows are slightly larger than other sparrows. They have brown backs, gray heads, and black-and-white stripes around the throat. These sparrows spend most of their time on the ground hiding in shrubs and looking for food. Instead of flying from place to place, they tend to scurry along the ground. They have a diet rich in seeds, with insects added during breeding season. During migration, they travel in flocks with other sparrow species. In the summer, the male Bell's sparrow sings a fast tune of trills and chirps.

HOW TO SPOT

Size: 5 to 6 inches (12.7 to 15.2 cm) long; 0.5 to 0.8 ounces (14.2 to 22.7 g)

Range: Southwestern United States and Baja California peninsula

Habitat: Chaparral, sagebrush, and shrublands

Diet: Insects, seeds, and fruits

BLACK-CHINNED SPARROW

(SPIZELLA ATROGULARIS)

Black-chinned sparrows have long tails. They have dark gray bodies with brown wings and backs. Breeding males have a small black patch on the throat and chin and sing loudly from perches. These sparrows breed on rocky hillsides and spend winters in desert scrublands. They look for insects on the ground and in shrubs and trees during breeding season. In winter, they eat seeds. Black-chinned sparrows living near the Mexican border don't always migrate.

FUN FACT

The song of this sparrow sounds like a dropped table-tennis ball. As it sings, the notes become faster and faster.

Male

Female

HOW TO SPOT

Size: 5.8 inches (14.7 cm) long; 0.3 to 0.5 ounces (8.5 to 14.2 g)

Range: Southwestern United States and Mexico

Habitat: Chaparral and sagebrush

Diet: Insects and seeds

BLACK-THROATED SPARROW

(AMPHISPIZA BILINEATA)

Black-throated sparrows are gray and brown. Their tails are darker colored with white spots. They have gray faces with two white stripes and a black triangle on the throat. These sparrows look for food near shrubs or cacti. They eat mostly insects during breeding season and seeds at other times. They hop on the ground when looking for food, taking short, low flights from time to time. The black-throated sparrow is a quiet bird that makes soft tinkling calls when perched in shrubs or trees.

HOW TO SPOT

Size: 4.7 to 5.5 inches (12 to 14 cm) long; 0.4 to 0.5 ounces (11.3 to 14.2 g)

Range: Mexico and western and southwestern United States

Habitat: Desert scrublands, canyons, and forests

Diet: Insects and seeds

BOTTERI'S SPARROW

(PEUCAEA BOTTERII)

Botteri's sparrows are rather large for sparrows. They have flat heads and long tails. Adults have mixed gray-and-brown backs, brown heads, and beige undersides. Young birds have deeper brown bodies with streaked beige undersides. Botteri's sparrows spend much of their time capturing insects on the ground and in the air. Often blending in with their surroundings, these birds are difficult to spot. Their music, however, draws attention in the summer when males sing from perches. Their song is a series of sharp and varied chirps followed by a fast trill.

HOW TO SPOT

Size: 5.9 to 6.7 inches (15 to 17 cm) long; 0.6 to 0.9 ounces (17 to 25.5 g)

Range: Southwestern United States and Mexico

Habitat: Grasslands and pastures

Diet: Insects and seeds

BREWER'S SPARROW

(SPIZELLA BREWERI)

Brewer's sparrows are slim birds with long tails and short, rounded wings. Males and females are different sizes, and sizes vary by region as well. These sparrows have gray undersides and throats, and they have gray-and-brown streaked backs and heads. These birds also have a white eye-ring. Brewer's sparrows are the most commonly found bird in the sagebrush of the western United States. They live in sagebrush shrubs during both breeding and wintertime. They pluck insects from tree bark and seeds from the ground. Males sing long trills in spring and summer.

HOW TO SPOT

Size: 5.1 to 5.9 inches (13 to 15 cm) long; 0.4 to 0.5 ounces (11.3 to 14.2 g)

Range: Southern Canada, United States, and Mexico

Habitat: Shrublands, woodlands, and grasslands

Diet: Insects and seeds

FUN FACT

These sparrows are well-adapted to dry habitats year-round. They can survive for weeks without drinking water.

CALIFORNIA TOWHEE

(MELOZONE CRISSALIS)

California towhees are large sparrows with short, rounded wings and long tails. Their light brown bodies have a reddish patch under the tail. These sparrows are often found in shrubby urban parks and yards. They eat mostly grass seeds, but they also eat insects and fruits during breeding season. To eat seeds, they take the grass stem in their mouths and slide off the seeds with their beaks. California towhees sing a bright song and often tap on windows, fighting their own reflection.

FUN FACT

California towhees rely on poison oak. They build their nests on this plant and eat its berries.

HOW TO SPOT

Size: 8.3 to 9.8 inches (21 to 25 cm) long; 1.3 to 2.4 ounces (36.9 to 68 g)
Range: California and southern Oregon
Habitat: Yards, chaparral, canyons, and urban parks
Diet: Insects, seeds, snails, and fruits

CASSIN'S SPARROW

(PEUCAEA CASSINII)

Cassin's sparrows are large sparrows with flat foreheads, big bills, and long, rounded tails. Adults have brown, gray, and red backs and heads. They have streaked undersides and a pale eye-ring. Young birds have beige undersides and no eye-ring. These sparrows live in dry grasslands or trees such as oak and acacia. They look for food on the ground in open spaces. Seeds are their main food in winter. Insects make up their diet at other times of the year. Cassin's sparrows sing a loud, melodious song.

FUN FACT

Male Cassin's sparrows perform when they sing. They fly straight up in the sky and float downward with their wings still.

HOW TO SPOT

Size: 5.1 to 5.9 inches (13 to 15 cm) long; 0.6 to 0.7 ounces (17 to 19.8 g)

Range: Southern United States and Mexico

Habitat: Grasslands and shrublands

Diet: Insects and seeds

CHIPPING SPARROW

(SPIZELLA PASSERINA)

Chipping sparrows are found all over North America. They are frequent visitors to parks, gardens, and backyards. These long-tailed sparrows change colors with the season and are slimmer than other sparrows. In summer, the chipping sparrow has a grayish back, a white underside, a red head, and a black line over the eye. In winter, its colors are paler with a darkly streaked back and a rufous head. They eat mostly seeds in fall and winter. Insects make up most of their diet in spring and summer. Chipping sparrows sing long trills that are easy to hear.

HOW TO SPOT

Size: 4.7 to 5.9 inches (12 to 15 cm) long; 0.4 to 0.6 ounces (11.3 to 17 g)

Range: Canada, United States, Mexico, Central America, and Caribbean

Habitat: Forests, woodlands, and urban parks

Diet: Insects, seeds, and spiders

MOLTING

Bird feathers wear out from flying, rubbing, and being exposed to the elements. Molting helps birds grow new feathers. The process varies by species, age, and time of year. Most songbirds molt once every summer, before fall migration. Some songbird species molt two or three times annually.

FIELD SPARROW

(SPIZELLA PUSILLA)

Field sparrows have pink bills and long tails. They are slimmer than most sparrows. They have brown-streaked backs and white chests. They have gray heads and a white eye-ring. These sparrows live in areas such as abandoned farm fields and openings in wooded areas. They eat mostly grass seeds in winter. Insects are added to their diet in spring. Their population is slowly declining as homes are built in their habitats. The field sparrow's song is a clear, bouncy trill heard across long distances.

HOW TO SPOT

Size: 4.7 to 5.9 inches (12 to 15 cm) long; 0.4 to 0.5 ounces (11.3 to 14.2 g)

Range: United States and northern Mexico

Habitat: Pastures, forest edges, and orchards

Diet: Seeds, insects, spiders, and snails

FOX SPARROW *(PASSERELLA ILIACA)*

Fox sparrows are big, round sparrows with medium-length tails. They have red-and-gray heads and reddish-brown backs. Their bills are yellow to dark gray. There are brown spots on the undersides and chests. These sparrows eat mostly insects during breeding season. Spiders, seeds, and fruits are also part of their diet in other seasons. Fox sparrows are most visible during winters spent in scrublands and forests. In summer, males sing a whistling song and make a smacking call.

HOW TO SPOT

Size: 5.9 to 7.5 inches (15 to 19 cm) long; 0.9 to 1.6 ounces (25.5 to 45.4 g)

Range: Canada and United States

Habitat: Forests, thickets, and chaparral

Diet: Insects, seeds, and fruits

FUN FACT

Fox sparrows find prey using a "double scratch" method. They hop forward and then back on the ground, scratching both feet backward through leaf litter.

VESPER SPARROW

(POOECETES GRAMINEUS)

Vesper sparrows are large, chubby sparrows. Their tails are long and notched. The body is brown with dark streaks, and they have a white eye-ring. The wings are tipped in white, and a chestnut-brown patch is hidden on the shoulder. These sparrows are usually found on the ground, hopping in grass and looking for food. They live in open grassland habitats, including prairies, pastures, and mountain meadows. Vesper sparrows sing a sweet tinkling song all day and just past sunset. The song is marked by whistles and trills.

HOW TO SPOT

Size: 5.1 to 6.3 inches (13 to 16 cm) long; 0.7 to 1 ounce (19.8 to 28.3 g)

Range: Southern Canada, United States, Mexico

Habitat: Shrublands and grasslands

Diet: Insects, spiders, and seeds

AMERICAN ROBIN

(TURDUS MIGRATORIUS)

American robins are the largest North American thrushes. The legs and tails are long. They have gray-and-brown backs, black heads, and orange undersides. A white patch on the belly becomes visible in flight. Females have paler heads than males. American robins are very common across North America. They visit city parks and backyard birdfeeders and are often seen pulling earthworms from soil. These cheerful birds sing a whistling song. Males attract females by singing, raising and spreading their tails, shaking their wings, and puffing out their throats.

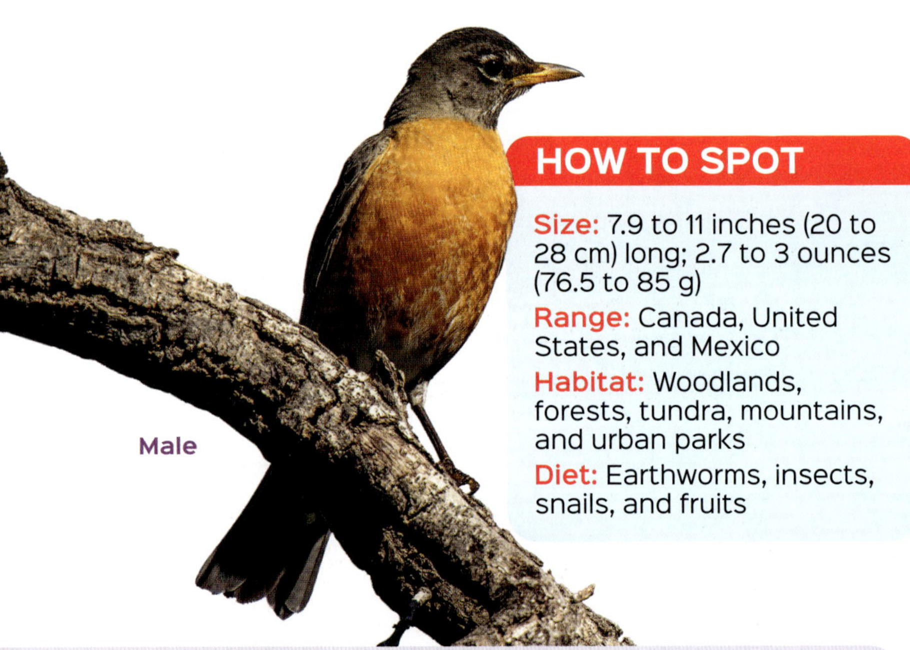

Male

HOW TO SPOT

Size: 7.9 to 11 inches (20 to 28 cm) long; 2.7 to 3 ounces (76.5 to 85 g)

Range: Canada, United States, and Mexico

Habitat: Woodlands, forests, tundra, mountains, and urban parks

Diet: Earthworms, insects, snails, and fruits

WHAT ARE THRUSHES?

More than two dozen thrushes are found in North America. These songbirds are small- to medium-sized birds. The lower leg is booted, or covered with a single scale, and the bill is narrow. Thrushes molt once a year. Northern species often migrate in the fall, returning home in the spring.

BICKNELL'S THRUSH

(CATHARUS BICKNELLI)

Bicknell's thrushes are small thrushes. Their tails are medium length. They have a light brown back and a gray face. Their undersides are light gray and brown with dark brown spots on the throat and chest. These thrushes often nest in stunted forests of balsam fir and spruce fir. They live near dead trees. They feed mostly near the ground, scratching for food or waiting to find it. Their song is high-pitched and flute-like, and it has a stuttering quality. Because of their small range and climate change, their population is declining.

HOW TO SPOT

Size: 6.3 to 6.7 inches (16 to 17 cm) long; 0.9 to 1.1 ounces (25.5 to 31.2 g)

Range: Eastern United States, Puerto Rico, and Cuba

Habitat: Stunted forests

Diet: Insects, ants, beetles, and fruits

EASTERN BLUEBIRD *(SIALIA SIALIS)*

Eastern bluebirds are small thrushes with a big head, big eyes, and chubby bodies. They have blue backs and heads with red throats and red-and-white undersides. Females have more gray on the back. These thrushes nest in cavities such as birdhouses and human-made nests. They perch on fences or wires in pairs or small flocks. Eastern bluebirds catch insects on the ground year-round, adding large amounts of fruit to the diet in fall and winter. These musical birds have a short and shaky song.

Female

Male

FUN FACT

The eastern bluebird is the official state bird of both New York and Missouri.

HOW TO SPOT

Size: 6.3 to 8.3 inches (16 to 21 cm) long; 1 to 1.1 ounces (28.3 to 31.2 g)

Range: Southern Canada, United States, Mexico, and Central America

Habitat: Pastures, yards, and urban parks

Diet: Insects and fruits

GRAY-CHEEKED THRUSH

(CATHARUS MINIMUS)

Gray-cheeked thrushes are medium-sized thrushes with long wings. They have gray-and-brown backs and heads, gray undersides, and dark spots on the chest and throat. The face is plain with an incomplete light gray eye-ring. These thrushes breed in thickets and dense undergrowth of different tree species. During breeding season, they search for insect prey by hopping on the ground and grabbing them quickly. They eat large amounts of berries and other fruits during winter and migration. They sing a flute-like song while perched high on fences.

HOW TO SPOT

Size: 6.3 to 6.7 inches (16 to 17 cm) long; 0.9 to 1.1 ounces (25.5 to 31.2 g)

Range: Canada, United States, Mexico, and Central America

Habitat: Boreal forests, thickets, and woodlands

Diet: Insects, spiders, and fruits

HERMIT THRUSH

(CATHARUS GUTTATUS)

Hermit thrushes are small thrushes with round heads and long tails. The back and head are brown, and the tail is red. They have beige undersides with dark spots on the throat and chest. The hermit thrush can be found in forests and open areas such as pond edges and trails. These birds eat mostly insects in spring, with the occasional small reptile or amphibian. Hermit thrushes sometimes collect insects by shaking grass to release the prey. In winter, more fruit is added to the diet. These thrushes sing a sweet, sad melody.

FUN FACT

Hermit thrushes are the only North American thrush to winter in the United States.

HOW TO SPOT

Size: 5.5 to 7.1 inches (14 to 18 cm) long; 0.8 to 1.3 ounces (22.7 to 36.9 g)

Range: Canada, United States, and Mexico

Habitat: Boreal forests, deciduous forests, and mountain forests

Diet: Insects, fruits, amphibians, and reptiles

SWAINSON'S THRUSH

(CATHARUS USTULATUS)

Swainson's thrushes are medium-sized thrushes with round heads, long wings, and medium-length tails. They have light brown backs and heads, brown-spotted undersides, and beige eye-rings. The throat is beige with a brown stripe on each side. Swainson's thrushes eat mostly insects during breeding season, adding fruits in other seasons. They look for food while perched on tree branches. These thrushes breed mainly in coniferous forests, except for coastal California species that prefer woodlands and thickets. Swainson's thrushes are musical birds with a flute-like song.

HOW TO SPOT

Size: 6.3 to 7.5 inches (16 to 19 cm) long; 0.8 to 1.6 ounces (22.7 to 45.4 g)

Range: Canada, United States, Mexico, and Central America

Habitat: Coniferous forests, woodlands, and thickets

Diet: Insects and fruits

FUN FACT

Singing Swainson's thrushes are tough to spot. Their softer songs sound like they come from a distant location.

VARIED THRUSH

(IXOREUS NAEVIUS)

Varied thrushes have large heads and long legs. They have dark-blue-and-gray backs. Their undersides, necks, and chests are orange. They have a black band across the chest and an orange line over the eye. Females' backs are paler in color than males. In summer, these thrushes search in dead leaves for insects to eat. They eat seeds and fruits in winter. Varied thrushes make a ringing sound that echoes. They also make long, eerie whistling sounds during breeding season.

Male

HOW TO SPOT

Size: 7.5 to 10.2 inches (19 to 26 cm) long; 2.3 to 3.5 ounces (65.2 to 99.2 g)

Range: Canada and western United States

Habitat: Forests, urban parks, gardens, and stream corridors

Diet: Insects and fruits

Stream corridor

WOOD THRUSH

(HYLOCICHLA MUSTELINA)

Wood thrushes have a large head with a white eye-ring. They have a potbelly and a short tail. The back and head are rufous with black spots on the undersides. The colors of a wood thrush's body provide camouflage as it looks for prey in forest branches. They eat mostly insects in the summer. In the fall, these thrushes add fruits such as black cherries and blueberries to their diet to gain strength for migration. The wood thrush's song is clear and flute-like.

HOW TO SPOT

Size: 7.5 to 8.3 inches (19 to 21 cm) long; 1.4 to 1.8 ounces (39.7 to 51 g)

Range: United States, Mexico, Central America, and Cuba

Habitat: Deciduous forests, mixed forests, and forest edges

Diet: Insects and fruits

FUN FACT

Wood thrushes have a call that sounds like gunshots. It warns other wood thrushes about nearby predators.

ALTAMIRA ORIOLE

(ICTERUS GULARIS)

Altamira orioles are large orioles with long tails and chunky bodies. Adults have black bodies with orange heads and undersides. Young birds have yellow heads and bodies with olive-green backs. These orioles have a small range in the United States, but they are found in southern Texas year-round. They often look for fruits, insects, and larvae in tree branches. These orioles drink nectar by poking holes in flowers. Altamira orioles sing a sweet whistling song and make clucking and chatter sounds.

HOW TO SPOT

Size: 8.3 to 9.8 inches (21 to 25 cm) long; 1.7 to 2.3 ounces (48.2 to 65.2 g)

Range: Southern Texas, Mexico, and Central America

Habitat: Woodlands, farms, stream corridors, and orchards

Diet: Insects, larvae, and fruits

WHAT ARE BLACKBIRDS AND ORIOLES?

Blackbirds and orioles belong to the *Icteridae* family. They have pointed, cone-shaped bills and long, pointed wings. Females are smaller and less colorful than males. They look for insects and seeds on the ground. North American species migrate toward Central and South America in winter and return home in spring.

BALTIMORE ORIOLE

(ICTERUS GALBULA)

Baltimore orioles are medium-sized birds with thick necks and long legs. Adult males are black and orange with a black head and a white bar on each black wing. Females and young males have gray heads and backs, yellow-and-orange undersides, and two white wing bars. Baltimore orioles are especially common in the eastern United States. They sometimes search for insects high up in trees. They will eat nectar and fruits from backyard feeders. Baltimore orioles sing clear whistling songs. They also chatter frequently with a quick *chirp* sound.

Adult male

HOW TO SPOT

Size: 6.7 to 7.5 inches (17 to 19 cm) long; 1.1 to 1.4 ounces (31.2 to 39.7 g)

Range: Canada, United States, Mexico, Central America, and Cuba

Habitat: Woodlands, forest edges, orchards, and deciduous forests

Diet: Fruits and insects

BREWER'S BLACKBIRD

(EUPHAGUS CYANOCEPHALUS)

Brewer's blackbirds are small, long-legged birds with a long tail and a round head. Males have yellow eyes and shiny black bodies with hints of blue and green. Females have dark eyes and brown bodies with darker brown wings and tails. Brewer's blackbirds are often seen in residential areas such as city parks, lawns, and golf courses. These blackbirds sometimes pick insects off farm animals or catch them midair. They occasionally eat baby birds, small frogs, and young voles. Their song is a loud chuckling sound often heard from treetops.

HOW TO SPOT

Size: 7.9 to 9.8 inches (20 to 25 cm) long; 1.8 to 3 ounces (51 to 85 g)

Range: Southern Canada, United States, and Mexico

Habitat: Marshes, grasslands, woodlands, chaparral, and sagebrush

Diet: Seeds, grains, and insects

Male

BROWN-HEADED COWBIRD

(MOLOTHRUS ATER)

Brown-headed cowbirds have a short tail and a thick head. The bill is shorter and thicker than those of other blackbirds. Males have shiny black bodies and brown heads. Females are simple brown birds with paler heads. These birds live in grasslands with low and scattered trees. They also live in thickets and residential areas. In winter, they spend time with other blackbird species in large flocks. These birds often destroy the eggs and babies of other songbirds. Males make a gurgling sound, while females chatter.

Male

Female

HOW TO SPOT

Size: 6.3 to 8.7 inches (16 to 22 cm) long; 1.3 to 1.8 ounces (36.9 to 51 g)

Range: Canada, United States, Mexico, and Central America

Habitat: Grasslands, thickets, pastures, orchards, and yards

Diet: Seeds, grains, and insects

BULLOCK'S ORIOLE

(ICTERUS BULLOCKII)

Bullock's orioles are medium-sized orioles with slender bodies and long tails. Adult males have orange faces and black backs and throats. Their wings have large white patches. Females and young birds have yellow-and-orange heads and tails, gray backs, and white-edged wings. These orioles often grab insects and fruits from trees or bushes. Using a method called gaping, they insert their closed bills inside fruit and caterpillars to suck up the juices. These orioles also remove stingers from bees before eating them.

FUN FACT

Bullock's orioles are very noisy birds. They sing sweet whistling songs and make harsh chatter sounds.

Adult male

HOW TO SPOT

Size: 6.7 to 7.5 inches (17 to 19 cm) long; 1 to 1.5 ounces (28.3 to 42.5 g)

Range: United States, Mexico, and Central America

Habitat: Woodlands, forests, urban parks, and stream corridors

Diet: Fruits, nectar, and insects

HOODED ORIOLE

(ICTERUS CUCULLATUS)

Hooded orioles have long bodies, rounded tails, and downward-curved bills. Adult males have black bodies with yellow or orange heads and white wing bars. Females have olive-and-yellow bodies with gray backs and white wing bars. Young males look like females with black throats. These orioles sometimes hang upside down while snatching prey. They eat fruits and drink nectar from flowering plants and backyard feeders. Hooded orioles are very talkative birds that cry out and sing cheerful tunes.

Adult male

FUN FACT

Females stitch hanging nests to the undersides of palm leaves. For this reason, these birds are sometimes called palm-leaf orioles.

HOW TO SPOT

Size: 7.1 to 7.9 inches (18 to 20 cm) long; 0.8 ounces (22.7 g)

Range: Southwestern United States, Mexico, and Central America

Habitat: Woodlands and scrublands

Diet: Insects, spiders, and fruits

ORCHARD ORIOLE *(ICTERUS SPURIUS)*

Orchard orioles have rounded heads and medium-length tails. Adult males have black backs and red undersides, a black throat and head, and a red wing patch. Females are green and yellow with two white wing bars. Young males resemble females with some black around the throat and bill. Orchard orioles pluck insects from treetops and drink flower nectar. They eat mostly pollen, fruits, and nectar during fall and winter. Males chatter and sing a whistling song to attract females.

HOW TO SPOT

Size: 5.9 to 7.1 inches (15 to 18 cm) long; 0.6 to 1 ounces (17 to 28.3 g)

Range: Southern Canada, United States, Mexico, Central America, and Cuba

Habitat: Stream corridors, marshes, and thickets

Diet: Insects, fruits, and nectar

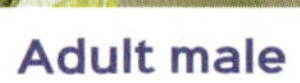

Adult male

WETLANDS

Wetlands such as swamps and marshes attract songbirds for different purposes. Many songbirds seek water and rely on the insects and plants that live and grow in these environments. Breeding, nesting, and raising young become easier with water close by. Wetlands also provide shelter and protection from land-based predators.

SCOTT'S ORIOLE *(ICTERUS PARISORUM)*

Scott's orioles are medium-sized orioles with long tails and sturdy legs. Males have a black back, chest, and throat with bright yellow undersides and white wing bars. Females have an olive-green back, chest, and throat with pale yellow undersides and pale white wing bars. Scott's orioles live in dry habitats near junipers, palms, and pinyon pines. They especially like to be near yucca. They search for insects on yucca plants, drink nectar from yucca flowers, and make nests woven from yucca leaf fibers.

FUN FACT

Scott's orioles are active singers. Their songs start before sunrise and last all day long at any time of year.

Male

HOW TO SPOT

Size: 9.1 inches (23 cm) long; 1.1 to 1.4 ounces (31.2 to 39.7 g)

Range: Southwestern United States and Mexico

Habitat: Scrublands, mountains, and deserts

Diet: Insects, fruits, and nectar

AMERICAN GOLDFINCH

(SPINUS TRISTIS)

American goldfinches have long wings. Males have yellow bodies, black foreheads and wings, and white wing bars. Females have olive-green bodies, pale yellow undersides, and white wing bars. In winter, all American goldfinches are dull brown with black wings and pale wing bars. Unlike other goldfinch subspecies, they molt twice a year, in late winter and late summer. The American goldfinch is often found in thistle plants and in backyards. These musical birds pair up with their mates in flight, mimicking each other's sounds.

HOW TO SPOT

Size: 4.3 to 5.1 inches (11 to 13 cm) long; 0.4 to 0.7 ounces (11.3 to 19.8 g)

Range: Southern Canada, United States, and Mexico

Habitat: Forest edges, woodlands, urban parks, and yards

Diet: Grass and tree seeds

Male

WHAT ARE FINCHES?

Finches are small and compact songbirds. There are hundreds of finch species, 17 of which live in North America. Finches have cone-shaped bills ideal for cracking seeds open. Their wings are slightly pointed, and their tails are notched. Finches have irregular fall migration patterns, with some staying close to home all year.

COMMON REDPOLL

(ACANTHIS FLAMMEA)

Common redpolls have brown-and-white bodies with streaked sides and two white wing bars. They have a black face with a red patch on the forehead. Males also have a pink-and-red patch on the chest. These finches are found in colder climates. They are a nomadic species that occasionally moves south in winter. They move wherever food is found, flying in large flocks. Seeds make up most of their winter diet. Insects, spiders, and fruits are added at other times of the year. These finches make calls with a zapping sound.

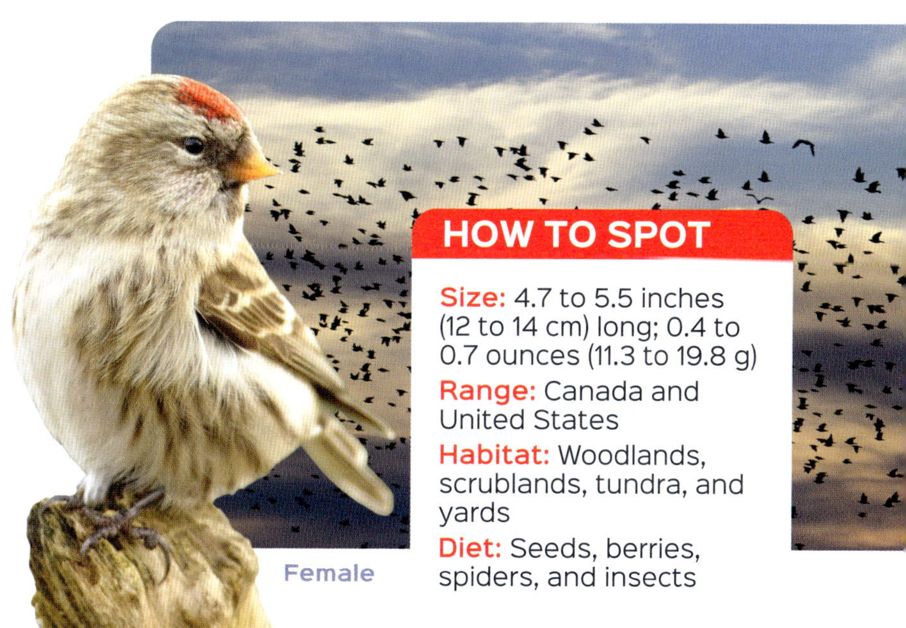

Female

HOW TO SPOT

Size: 4.7 to 5.5 inches (12 to 14 cm) long; 0.4 to 0.7 ounces (11.3 to 19.8 g)

Range: Canada and United States

Habitat: Woodlands, scrublands, tundra, and yards

Diet: Seeds, berries, spiders, and insects

IRRUPTIVE MIGRATION

Irruptive migration is when birds do not follow their usual migration pattern. This happens when the food supply in a bird's normal habitat is too low. Birds fly farther, looking for food. They may fly short distances. Sometimes large numbers of a species may fly very far south.

EVENING GROSBEAK

(COCCOTHRAUSTES VESPERTINUS)

Evening grosbeaks are large finches with a thick neck, big chest, and short tail. Adult males have yellow-and-black bodies with white wing patches. There are yellow stripes over the eyes. Females and young birds are gray with black-and-white wings and green-and-yellow necks. Evening grosbeaks use their giant bills to crush seeds that other finches cannot open. They also enjoy insects, fruits, and maple sap. The bird's call sounds vary from sharp to sweet.

HOW TO SPOT

Size: 6.3 to 7.1 inches (16 to 18 cm) long; 1.9 to 2.6 ounces (53.9 to 73.7 g)

Range: Canada, United States, and Mexico

Habitat: Coniferous forests, deciduous forests, orchards, and parks

Diet: Insects, larvae, fruits, and seeds

Male

HABITAT LOSS

Habitat loss is a serious problem for many songbirds. Many bird species are losing their homes because of human-related activities, including agriculture, deforestation, mining, and construction. Natural threats such as wildfires, weather events, disease, and predators are also reducing songbird populations.

HOARY REDPOLL

(ACANTHIS HORNEMANNI)

Hoary redpolls have white bodies with gray-and-brown streaks. The wings are gray with white wing bars. Adults have a small red patch on the head. These birds have fluffy feathers that help them survive in very cold habitats. They breed in barren arctic tundra in scattered shrubs and stunted trees. These active birds often hang upside down from branches in flocks, picking out seeds or searching for insects. The hoary redpoll makes a variety of sounds, including nasal whistling and short trills.

HOW TO SPOT

Size: 5 to 6 inches (12.7 to 15.2 cm) long; 0.4 to 0.7 ounces (11.3 to 19.8 g)

Range: Canada and northern United States

Habitat: Tundra and stream corridors

Diet: Seeds, plants, insects, and spiders

FUN FACT

When hoary redpolls get hot, they pluck out feathers to cool off. These feathers grow back in just a few days.

HOUSE FINCH *(HAEMORHOUS MEXICANUS)*

House finches have long, flat heads and short wings. Males have a rosy red face with brown streaks on the back, tail, and undersides. Females have no pink on the face. Their bodies are streaked with gray and brown. In the 1940s, these native birds of Mexico and the western United States were sold in the eastern United States. When people would not buy them, the birds were released from cages. They quickly populated the eastern coast up to southern Canada. House finches often visit backyard feeders and are found in large groups.

HOW TO SPOT

Size: 5.1 to 5.5 inches (13 to 14 cm) long; 0.6 to 0.9 ounces (17 to 25.5 g)

Range: Southern Canada, United States, Mexico, and Central America

Habitat: Grasslands, deserts, chaparral, forests, and yards

Diet: Seeds, buds, and fruits

Male

Female

FUN FACT

House finches make calls with a short *cheep* sound. They sing long warbling songs.

LAWRENCE'S GOLDFINCH

(SPINUS LAWRENCEI)

Lawrence's goldfinches have wide wings. Breeding adult males are gray with a yellow chest, wings, and back. Nonbreeding adult males and females have brown bodies and less yellow on the chest. Lawrence's goldfinches are nomadic birds. Their population during non-migratory periods can shift greatly in a region from one year to the next. They join other bird species in mixed flocks during migration and winter. The Lawrence's goldfinch will perch in plants to extract seeds while hanging upside down. These finches sing long, high-pitched trills.

Adult male

HOW TO SPOT

Size: 3.9 to 4.7 inches (10 to 12 cm) long; 0.3 to 0.5 ounces (8.5 to 14.2 g)

Range: Western and southwestern United States

Habitat: Chaparral, woodlands, farms, and stream corridors

Diet: Seeds, plants, and fruits

LESSER GOLDFINCH

(SPINUS PSALTRIA)

Lesser goldfinches have long wings and short tails. Males have shiny black or pale green backs and shiny black heads. Their undersides are yellow, and they have a white patch on the wing. Females and young birds have olive-green backs, pale yellow undersides, and black wings with two white wing bars. These finches live in cottonwood, willow, and scrubby oak habitats. They also visit backyards for water and seeds. Occasionally, they eat fruits and tree buds. Lesser goldfinches make light, wheezy sounds that often imitate other birds.

Adult male

HOW TO SPOT

Size: 3.5 to 4.3 inches (9 to 11 cm) long; 0.3 to 0.4 ounces (8.5 to 11.3 g)

Range: United States, Mexico, and Central America

Habitat: Thickets, woodlands, and farms

Diet: Seeds, fruits, and plants

PINE GROSBEAK

(PINICOLA ENUCLEATOR)

Pine grosbeaks are large finches with a big bill and a round head. Males are pink, red, and gray. Females and young birds are gray with hints of red, orange, or yellow on the head. All birds have dark gray wings with two white wing bars. Pine grosbeaks live in the same region year-round but may travel south to find more food. In addition to seeds, they sometimes eat insects and spiders in the summer. These finches drink water or eat snow every day and visit bird feeders that offer black oil sunflower seeds. Pine grosbeaks make a rich warbling sound.

Evergreen forest

Female

FUN FACT

Young birds cannot eat plants easily. Adults spit up a vegetable and insect paste stored in their jaw pouch and feed it to their young.

HOW TO SPOT

Size: 8 to 10 inches (20.3 to 25.4 cm) long; 2 to 2.8 ounces (56.7 to 79.4 g)

Range: Canada and United States

Habitat: Valleys and evergreen forests

Diet: Seeds, fruits, and buds

PINE SISKIN

(SPINUS PINUS)

Pine siskins are streaked brown finches with yellow edges on the wings and tail. They nest in coniferous and mixed forests. When searching for food, they travel to many habitats, including thickets, deciduous forests, grasslands, and residential lawns or yards. Pine siskins hang upside down from tree branches when picking seeds. They are very social birds that fly in flocks and chatter constantly when looking for food. They make a loud zipping or buzzing sound.

HOW TO SPOT

Size: 4.3 to 5.5 inches (11 to 14 cm) long; 0.4 to 0.6 ounces (11.3 to 17 g)

Range: Canada, United States, Mexico, and Central America

Habitat: Coniferous forests and mixed forests

Diet: Seeds, plants, and insects

PURPLE FINCH

(HAEMORHOUS PURPUREUS)

Purple finches are large finches. Males have a pink-and-red head and chest. Their backs are brown, and their undersides are white. Females have brown-streaked bodies. They have white stripes by the eyes and dark lines on the throat. In addition to conifer seeds, these finches eat berries and insects and also drink nectar. They spend summers in cool evergreen forests or near streams. They live in a larger variety of habitats at other times of the year. These musical birds sing a warbling song.

Male

Female

HOW TO SPOT

Size: 4.7 to 6.3 inches (12 to 16 cm) long; 0.6 to 1.1 ounces (17 to 31.2 g)

Range: Canada and United States

Habitat: Coniferous forests, mixed forests, and shrublands

Diet: Seeds, fruits, nectar, and insects

RED CROSSBILL

(LOXIA CURVIROSTRA)

Red crossbills are medium-sized birds with a twisted bill that crosses when closed. Males have red bodies with a darker red tail and wings. Females have olive-green or brown bodies with yellow undersides. They live in coniferous forests—especially pine, spruce, and Douglas fir. Red crossbills generally do not migrate but are very nomadic in winter. They travel wherever they find food and may show up far south of their normal range. These social birds sing trills or smooth warbling songs all year long.

FUN FACT

There are several types of red crossbills. Each has its own call and favorite conifer.

Male

HOW TO SPOT

Size: 5.5 to 7.5 inches (14 to 19 cm) long; 1.4 ounces (39.7 g)
Range: Canada, United States, Mexico, and Central America
Habitat: Coniferous forests
Diet: Conifer seeds, buds, berries, and insects

WHITE-WINGED CROSSBILL

(LOXIA LEUCOPTERA)

White-winged crossbills are medium-sized finches with forked tails, crossed bills, and two white wing bars. Adult males are pink with a black tail and wings. Females and young male birds are yellow with brown streaks. The nomadic white-winged crossbill can breed at any time of year, and it remains in flocks all year, even during breeding season. This species searches for food in coniferous forests but also in weedy fields or pine forests. When food is scarce, white-winged crossbills may wander far from their normal range. They make rattling and high-pitched trill sounds.

HOW TO SPOT

Size: 5.9 to 6.7 inches (15 to 17 cm) long; 0.8 to 0.9 ounces (22.7 to 25.5 g)

Range: Canada and United States

Habitat: Coniferous forests

Diet: Seeds, insects, and spiders

Adult male

FUN FACT

These birds can eat up to 3,000 conifer seeds per day.

BLACK-HEADED GROSBEAK

(PHEUCTICUS MELANOCEPHALUS)

Black-headed grosbeaks have large cone-shaped bills, short tails, and short, thick necks. Adult males have black heads and orange-and-brown bodies. Their wings are black and white. Females and young males have brown backs, and their chests are pale orange. All black-headed grosbeaks have gray bills. They use their giant bills to crack open seeds and crush insects they find in trees. They also eat fruits such as figs, cherries, and apricots. Black-headed grosbeaks sing a rich whistling song that slowly drops in tone.

FUN FACT
Both males and females sit on eggs and feed their babies.

Adult male

HOW TO SPOT

Size: 7.1 to 7.5 inches (18 to 19 cm) long; 1.2 to 1.7 ounces (34 to 48.2 g)

Range: United States, Mexico, and southern and western Canada

Habitat: Canyons, valleys, thickets, yards, and orchards

Diet: Insects, seeds, and fruits

WHAT ARE CARDINALS?

Cardinals are medium-sized songbirds found in North, Central, and South America. They include some species of grosbeaks and tanagers. Cardinals have thick bills and are known for their clear whistling song. Some cardinals migrate farther south in the fall and return home in the spring.

BLUE GROSBEAK

(PASSERINA CAERULEA)

Blue grosbeaks are large and stocky cardinals. The male blue grosbeak is deep blue with small black masks around the eyes. Females are deep brown with blue tails. Both sexes have a light brown-and-gray wing bar. These cardinals breed near roads and other open areas. They build nests in shrubs, small trees, and vines. Blue grosbeaks search for food on the ground and in trees. They eat crickets and grasshoppers, as well as seeds. These birds sing rich warbling songs from trees and perches.

HOW TO SPOT

Size: 5.9 to 6.3 inches (15 to 16 cm) long; 0.9 to 1.1 ounces (25.5 to 31.2 g)

Range: United States, Mexico, Central America, Cuba, and Puerto Rico

Habitat: Shrublands, forest edges, deserts, and stream corridors

Diet: Insects, snails, and seeds

Male

FUN FACT

Blue grosbeaks often raise two broods of nestlings in a single breeding season.

CLASSIFYING GROSBEAKS

Grosbeaks have one major feature in common: their thick, cone-shaped bills. But the similarities sometimes stop there. Some are classified as finches, while others belong to the cardinal family. They are grouped differently because of major differences in other features, habitat, and behavior.

LAZULI BUNTING

(PASSERINA AMOENA)

The lazuli bunting is a small, stocky cardinal with a sloped forehead and a notched tail. Adult breeding males have blue backs and orange chests with white undersides. Females have gray-and-brown backs with a hint of blue in the wings and tail, two beige wing bars, and a tan chest. Young birds have orange chests with blue-and-tan backs and heads. They pluck insects and seeds from grasses and leaves. They also pick berries off bushes.

HOW TO SPOT

Size: 5.1 to 5.9 inches (13 to 15 cm) long; 0.5 to 0.6 ounces (14.2 to 17 g)

Range: Southern Canada, United States, Mexico, and Cuba

Habitat: Chaparral, thickets, brushlands, and scrub oaks

Diet: Insects, berries, and seeds

Adult male

FUN FACT

Males sing squeaky, muffled songs from shrubs to defend their territory.

NORTHERN CARDINAL

(CARDINALIS CARDINALIS)

Northern cardinals are large members of the cardinal family. They have a long tail and a large crest. Males are bright red all over, with a black face surrounding a red bill. Females are pale brown with red highlights in the tail, wings, and crest. They have the same face markings as males. Northern cardinals nest in dense leaves and sing on high, visible perches. They eat mostly seeds and fruits with some insects on the side. These birds do not migrate. Northern cardinals make a sweet whistling sound, especially in the summer.

Male

FUN FACT

The northern cardinal's feathers remain the same color year-round, making it stand out in the snow.

HOW TO SPOT

Size: 8.3 to 9.1 inches (21 to 23 cm) long; 1.5 to 1.7 ounces (42.5 to 48.2 g)

Range: Southeastern Canada, United States, Mexico, and Central America

Habitat: Thickets, yards, and forest edges

Diet: Seeds, fruits, and insects

PAINTED BUNTING *(PASSERINA CIRIS)*

Painted buntings vary in color. Males have bright blue heads, red undersides, and yellow-green backs. Females and young birds have yellow-and-green bodies with pale eye-rings. Painted buntings in the southeastern United States breed in scrublands, thickets, yards, and citrus groves. Others may breed in abandoned farms and patches of weeds, grasses, and wildflowers. Painted buntings eat seeds most of the year. They add insects during breeding season. These birds sing gentle, rambling songs.

Adult male

Female

HOW TO SPOT

Size: 4.7 to 5.1 inches (12 to 13 cm) long; 0.6 ounces (17 g)
Range: United States, Mexico, Central America, and Cuba
Habitat: Thickets, rain forests, and savannas
Diet: Insects and seeds

ROSE-BREASTED GROSBEAK

(PHEUCTICUS LUDOVICIANUS)

Rose-breasted grosbeaks have wide chests, short necks, and square-shaped tails. Males have black-and-white bodies. A bright red triangle extends from the black throat to the chest. Females and young birds have brown streaks with white stripes over the eye. These cardinals eat many insects, fruits, and seeds during breeding season. Berries are added during fall migration. The male rose-breasted grosbeak's song defines territory and attracts females. Both males and females sing sweet, whistled songs and make short *chink* calls.

HOW TO SPOT

Size: 7.1 to 8.3 inches (18 to 21 cm) long; 1.4 to 1.7 ounces (39.7 to 48.2 g)

Range: Canada, United States, Mexico, Central America, and Cuba

Habitat: Deciduous forests, mixed forests, and thickets

Diet: Seeds, insects, and fruits

Adult male

Female

SCARLET TANAGER

(PIRANGA OLIVACEA)

Scarlet tanagers have stocky bodies and thick, rounded bills. Their heads are big, and their tails are short and broad. In spring and summer, males are bright red with black wings and tails. Females and young birds are olive-green and yellow. After breeding, adult male feathers turn olive-green and yellow while the wings and tail remain black. Scarlet tanagers eat mostly insects in the summer but add fruits during migration and winter. They hover with fast wingbeats to grab insects. These cardinals make whirring and harsh call sounds.

HOW TO SPOT

Size: 6.3 to 6.7 inches (16 to 17 cm) long; 0.8 to 1.3 ounces (22.7 to 36.9 g)

Range: Southern Canada, United States, Mexico, Central America, Cuba, and Puerto Rico

Habitat: Deciduous forests, mixed forests, shrublands, and yards

Diet: Insects and fruits

Breeding male

SUMMER TANAGER

(PIRANGA RUBRA)

Summer tanagers are chunky birds with large bodies and heads. Adult males are bright red all over. Females and young males are bright yellow and green. They have yellow heads and undersides. Their backs and wings are green. These birds are the most commonly found tanager in North America. They eat a wide variety of insects and fruits. The male summer tanager makes slurred whistling sounds at other males to defend its territory.

Adult Male

HOW TO SPOT

Size: 6.7 inches (17 cm) long; 0.9 to 1.1 ounces (25.5 to 31.2 g)

Range: United States, Mexico, and Central America

Habitat: Deciduous forests, forest edges, and river edges

Diet: Insects, spiders, and fruits

FUN FACT

Bees and wasps are the summer tanager's favorite foods year-round.

CHESTNUT-COLLARED LONGSPUR *(CALCARIUS ORNATUS)*

Chestnut-collared longspurs have large heads and short tails. They tend to live in habitats with short grass. Adult breeding males have a beige face and throat and a brown neck. The chest and undersides are black. Females and nonbreeding males are pale gray with streaked chests. Chestnut-collared longspurs bob their heads up and down while walking or running on the ground. They eat grasshoppers and seeds picked off the ground. This longspur sings a sweet warbling song with a gurgling sound.

Breeding male

HOW TO SPOT

Size: 5 to 6.5 inches (12.7 to 16.5 cm) long; 0.6 to 0.8 ounces (17 to 22.7 g)

Range: Canada, United States, and Mexico

Habitat: Grasslands and deserts

Diet: Insects and seeds

WHAT ARE LONGSPURS?

Longspurs are songbirds of the *Calcariidae* family. Four species are found in North America. Longspurs are small birds with a distinctive long claw on each hind foot. These active birds fly often and rarely perch. They migrate short distances in the fall, staying within the continent.

LAPLAND LONGSPUR

(CALCARIUS LAPPONICUS)

Lapland longspurs have a big head, a short tail, and long wings. Adult breeding males have black faces. A rufous patch and a yellow-white line are found on the neck. Females have a similar look with less black. In winter, both sexes lose their bright colors. They become pale brown and streaked. These longspurs spend most of their time looking for seeds. Insects are added to the diet during breeding season. They breed in arctic or alpine tundra and spend winters in remote fields. Lapland longspurs sing a series of loud jingling and squeaky sounds.

Arctic tundra

Breeding male

FUN FACT

In the summer, an adult Lapland longspur can eat between 3,000 and 10,000 insects and seeds daily. Nestlings are fed an extra 3,000 insects per day.

HOW TO SPOT

Size: 5.9 to 6.3 inches (15 to 16 cm) long; 0.8 to 1.2 ounces (22.7 to 34 g)

Range: Canada and United States

Habitat: Arctic tundra and alpine tundra

Diet: Seeds and insects

SMITH'S LONGSPUR

(CALCARIUS PICTUS)

Smith's longspurs have a short, notched tail. The bill is long and cone-shaped. Breeding males have deep brown undersides with black-and-white stripes on the face. The back and wings are streaked with brown and beige. Breeding females have a similar look, with beige undersides. A Smith's longspur will remove the wings and legs of insects before eating them. Similarly, they remove seed husks before swallowing. Males and females both have many mates at once—an unusual trait for songbirds. These longspurs sing a sweet warbling song.

HOW TO SPOT

Size: 5.8 to 6.5 inches (14.7 to 16.5 cm) long; 1 ounce (28.3 g)

Range: Canada and United States

Habitat: Prairies, shrublands, and tundra

Diet: Insects, seeds, and spiders

THICK-BILLED LONGSPUR

(RHYNCHOPHANES MCCOWNII)

True to its name, the thick-billed longspur has a thick and cone-shaped bill. Breeding males have black-and-white patterned heads with a black bill, a black chest patch, and light brown wing patches. Nonbreeding males have a brown head with a pink bill, a slightly black chest patch, and slightly brown wing patches. Females and young birds have beige-and-brown streaks with a white throat and undersides. These longspurs search for seeds and insects on the ground. Sometimes they chase after insects. They sing a tinkling, warbling song.

FUN FACT

These longspurs remove the wings and legs from grasshoppers before feeding them to their young.

HOW TO SPOT

Size: 5.9 inches (15 cm) long; 0.8 ounces (22.7 g)

Range: Canada, United States, and Mexico

Habitat: Prairies, dry lake beds, and farms

Diet: Insects and seeds

Breeding male

BENDIRE'S THRASHER

(TOXOSTOMA BENDIREI)

Bendire's thrashers are medium-sized, slender birds. They have a pale brown back and head with paler undersides and a spotted chest. The tail feathers are gray. They spend most of their time on the ground searching for food. They use their bills to flip over rocks and plants in search of prey and dig them out of the ground. Bendire's thrashers also climb trees to search for fruits or insects. As a suitable habitat shrinks, their population declines. These songbirds sing a rich warbling song.

HOW TO SPOT

Size: 9 to 9.8 inches (23 to 25 cm) long; 2 to 2.4 ounces (56.7 to 68 g)

Range: Mexico and southwestern United States

Habitat: Deserts, farms, and shrublands

Diet: Insects, spiders, and fruits

WHAT ARE MOCKINGBIRDS AND THRASHERS?

Mockingbirds and thrashers are skilled at mimicking other bird songs. All 35 species are found in North, Central, and South America. The bill is thin and downcurved. They have short, rounded wings and long tails. Many mockingbirds and thrashers stay in their region year-round, though northern birds migrate south in the fall.

BROWN THRASHER

(TOXOSTOMA RUFUM)

Brown thrashers are large, slim birds with long and powerful legs. They have a gray-and-brown face with yellow eyes. They have brown, streaked bodies with white undersides. The wings have black-and-white wing bars. These thrashers look for food on the ground while hiding under dense cover. They catch insects in the air and pluck berries from bushes. Brown thrashers can sing more than 1,000 different song patterns. They also imitate the songs of other birds.

HOW TO SPOT

Size: 9.1 to 11.8 inches (23 to 30 cm) long; 2.1 to 3.1 ounces (59.5 to 87.9 g)

Range: United States and southern Canada

Habitat: Thickets, forest edges, and woodlands

Diet: Insects, fruits, seeds, and nuts

FUN FACT

Brown thrashers are very protective of their nests. They are known for attacking people and dogs that get too close.

CALIFORNIA THRASHER

(TOXOSTOMA REDIVIVUM)

California thrashers are large, slender birds with long legs and wide wings. They have gray-and-brown bodies with a beige throat. A beige or red patch is found under the tail. Most California thrashers spend their whole lives in their habitat, extending from southern Oregon to the Baja California peninsula. They eat mostly insects during breeding season, pouncing on moving prey. Fruits are added to the summer diet when insects are scarce. Both males and females are musical, often singing together from the tops of shrubs.

HOW TO SPOT

Size: 12.6 inches (32 cm) long; 2.8 to 3.3 ounces (79.4 to 93.6 g)

Range: California and southern Oregon

Habitat: Chaparral, scrub oaks, and sagebrush

Diet: Insects and fruits

CRISSAL THRASHER

(TOXOSTOMA CRISSALE)

Crissal thrashers are large birds with long legs. They have gray bodies with dark red streaks under the wings. These thrashers live in deserts and dry, brushy habitats, rarely straying from home. Nests are built under branches to shield them from desert heat and predators. Crissal thrashers search for prey on the ground, using their bills to pluck insects and dig holes. These thrashers are among the most musical desert songbirds, singing a gentle song.

HOW TO SPOT

Size: 11.8 inches (30 cm) long; 1.9 to 2.5 ounces (53.9 to 70.9 g)

Range: Mexico and southwestern United States

Habitat: Deserts, canyons, and brushlands

Diet: Insects, spiders, fruits, and seeds

FUN FACT

Crissal thrashers run and walk more than they fly. If a predator approaches, they run away.

CURVE-BILLED THRASHER

(TOXOSTOMA CURVIROSTRE)

Curve-billed thrashers are slender birds with long downcurved bills and thick legs. They have orange eyes and gray-and-brown heads and backs. The undersides are cream-colored with gray and brown speckles. These thrashers use their bills to sweep through soil and leaf litter for food. Their bills help them catch insects and build nests safely on spiny cacti. These birds run and hop on the ground, never leaving their habitat. They make a whistling sound.

HOW TO SPOT

Size: 10.6 to 11 inches (27 to 28 cm) long; 2.1 to 3.3 ounces (59.5 to 93.6 g)

Range: Mexico and southwestern United States

Habitat: Grasslands, brushlands, thickets, deserts, and canyons

Diet: Insects, spiders, snails, fruits, and seeds

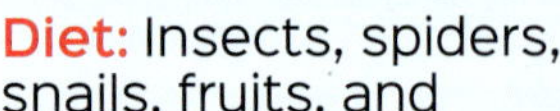

LeConte's Thrasher

(Toxostoma lecontei)

LeConte's thrashers are large songbirds with long, powerful legs. Their bodies are light brown with a darker brown tail and a dark line on the throat. They have a dark bill, eyes, and legs. These thrashers live in sandy deserts with scrublands where few other birds are found. They look for food in leaf litter by swiping with their large bills. They also chase after prey on foot and run away when predators approach. They have become less common due to habitat loss from agriculture and land development. The LeConte's thrasher makes high-pitched, squeaky calls.

HOW TO SPOT

Size: 9.6 to 11.4 inches (24.4 to 29 cm) long; 1.9 to 2.6 ounces (53.9 to 73.7 g)

Range: Southwestern United States

Habitat: Deserts

Diet: Insects, spiders, reptiles, and seeds

NORTHERN MOCKINGBIRD

(MIMUS POLYGLOTTOS)

Northern mockingbirds are medium-sized birds with a long tail. They have gray-and-brown bodies with a paler chest and underside. The wings have two white wing bars and a white patch. These birds do not migrate. They eat mostly insects in spring and summer, switching to mostly fruits in colder weather. Northern mockingbirds defend their territory and fight other birds that come close. Both males and females sing day and night. They use a variety of sounds such as whistles and trills, often mimicking other songbirds such as blackbirds, orioles, and jays.

HOW TO SPOT

Size: 8.3 to 10.2 inches (21 to 26 cm) long; 1.6 to 2 ounces (45.4 to 56.7 g)

Range: Southern Canada, United States, Mexico, Central America, Cuba, and Puerto Rico

Habitat: Thickets, shrublands, and yards

Diet: Insects and fruits

FUN FACT

Northern mockingbirds are known for imitating different sounds. They have been heard mimicking frogs, car alarms, and slamming doors.

SAGE THRASHER

(OREOSCOPTES MONTANUS)

Sage thrashers are the smallest thrashers. Their bills are short. They have a gray back and head with yellow eyes. Their undersides and chest are spotted, and they have two white wing bars. These thrashers breed only in sagebrush. Their sagebrush habitat is threatened by human development, livestock grazing, and invasive plants. During winter and migration, they live in open country with grasslands, bushes, and woodlands. They search for prey on the ground. Sage thrashers sing loud, cheerful songs and make chirp and whistle sounds.

FUN FACT

A male sage thrasher was once recorded singing for more than 22 minutes without a pause.

HOW TO SPOT

Size: 7.9 to 9.1 inches (20 to 23 cm) long; 1.4 to 1.8 ounces (39.7 to 51 g)

Range: Southwestern Canada, United States, and Mexico

Habitat: Sagebrush

Diet: Insects, ants, and fruits

GLOSSARY

boreal forest
A cold forest in the northern hemisphere, home to pine and spruce trees.

carrion
The flesh and bones of a dead animal.

chaparral
A thicket of dwarf evergreen oak trees.

coniferous forest
A forest consisting mainly of conifer trees, which are usually evergreen.

crest
A tuft on the head.

deciduous forest
A forest consisting mainly of trees that lose their leaves before winter.

deforestation
The removal of a wide area of trees.

larva
An insect that has hatched from its egg but is not yet an adult.

nomadic
Moving from place to place.

oasis
A fertile or green area in a desert region.

parasite
An organism that lives in, on, or with other organisms.

prey
An animal killed by another animal as food.

rufous
Reddish-brown in color.

syrinx
A vocal organ found in birds.

thicket
A dense growth of shrubs or small trees.

trill
A sound made by rapidly alternating tones.

tundra
Rolling, treeless plains in arctic and subarctic regions.

wing bar
A line of contrasting color on a bird's wing.

TO LEARN MORE

FURTHER READINGS

Earley, Chris G. *Warblers of Eastern North America.* Firefly Books, 2023.

Tekiela, Stan. *Attracting and Feeding Finches.* Adventure Publications, 2022.

Tekiela, Stan. *Attracting and Feeding Orioles.* Adventure Publications, 2022.

ONLINE RESOURCES

To learn more about songbirds, please visit **abdobooklinks.com** or scan this QR code. These links are routinely monitored and updated to provide the most current information available.

PHOTO CREDITS

Cover Photos: Agami Photo Agency/Shutterstock Images, front (California thrasher, yellow warbler); Danita Delimont/Shutterstock Images, front (lazuli bunting), back (curve-billed thrasher, painted bunting); FotoRequest/Shutterstock Images, front (blue jay); Janet Griffin/Shutterstock Images, front (Baltimore oriole); Jeff Rzepka/Shutterstock Images, front (blue warbler); Milan Zygmunt/Shutterstock Images, front (summer tanager); Mircea Costina/Shutterstock Images, front (perched goldfinch); Paul Reeves Photography/Shutterstock Images, front (blue-headed vireo); Spencer Bonynge/Shutterstock Images, front (thrush); Thomas Torget/Shutterstock Images, front (cardinal); Vlad G/Shutterstock Images, front (flying goldfinch); Marcin Perkowski/Shutterstock Images, back (raven)
Interior Photos: Agami Photo Agency/Shutterstock Images, 1 (top left), 5 (bottom right), 8 (top), 17 (bottom), 18, 19, 20 (left), 21, 22 (top), 55, 59 (top), 63, 65, 67, 75, 95, 98, 107; Brian A Wolf/Shutterstock Images, 1 (top right), 69; Coulanges/Shutterstock Images, 1 (center right); vagabond54/Shutterstock Images, 1 (bottom left), 96, 100, 105; Rob Palmer Photography/Shutterstock Images, 1 (bottom right), 5 (top right), 91; Brian Lasenby/Shutterstock Images, 4 (bottom left), 20 (right), 94; gergosz/Shutterstock Images, 4 (bottom right), 77 (left); Memno Schaefer/Shutterstock Images, 4 (top), 84; FotoRequest/Shutterstock Images, 5 (top left), 6, 40 (top), 44, 78; Harry Collins Photography/Shutterstock Images, 5 (bottom center), 89; Moab Republic/Shutterstock Images, 5 (bottom center), 12 (left); Paul Reeves Photography/Shutterstock Images, 5 (top center), 23 (right), 38 (right), 46; Kevin Manns/Shutterstock Images, 6 (top), 45 (top); fc56/Shutterstock Images, 6 (bottom), 45 (bottom); Tim Gray/Shutterstock Images, 8 (bottom); Thomas Torget/Getty Images, 9 (top); uttirat wiriyanon/Shutterstock Images, 9 (bottom); Mircea Costina/Shutterstock Images, 10 (top), 10 (bottom), 15 (bottom), 58; Mike Pellinni/Shutterstock Images, 11 (top), 11 (center); Gary W. Carter/Getty Images, 11 (bottom); Vineeth Radhakrishnan/Shutterstock Images, 12 (right), 81; Matthew Jolley/Shutterstock Images, 13; Ray Hennessy/Shutterstock Images, 14 (left); Jay Ondreicka/Shutterstock Images, 14 (right); pcnorth/Shutterstock Images, 15 (top); Barry and Carole Bowden/Shutterstock Images,16; JoshuaGoddard/Shutterstock Images, 17 (top); Pedro Bernal/Shutterstock Images, 18; Michelle Nyss/Shutterstock Images, 22 (bottom), 23 (left), 27 (left); punkbirdr/Shutterstock Images, 24; Rajh.Photography/Shutterstock Images, 25; Matt Cuda/Shutterstock Images, 26; Damsea/Shutterstock Images, 27 (right); A. Viduetsky/Shutterstock Images, 28; Nelson Sirlin/Shutterstock Images, 29; Sean R. Stubbern/Shutterstock Images,

30; Raul Baena/Shutterstock Images, 31; Christopher Unsworth/ Shutterstock Images, 32 (left); FJAH/Shutterstock Images, 32 (right); Michael Chatt/Shutterstock Images, 33; Hayley Crews/ Shutterstock Images, 34, 43 (top); Michael Woodruff/Shutterstock Images, 35 (top); Travis Maher/ Shutterstock Images, 35 (bottom), 53; Pablo Rodriguez Merkel/ Shutterstock Images, 36; Wim Hoek/Shutterstock Images, 37; 2009fotofriends/Shutterstock Images, 38 (left); Eivor Kuchta/ Shutterstock Images, 39 (top); Eivor Kuchta/Shutterstock Images, 39 (bottom); Tom Reichner/Shutterstock Images, 40 (bottom); Piotr Krzeslak/ Shutterstock Images, 41; Griffin Gillespie/Shutterstock Images, 42; Stas Moroz/Shutterstock Images, 43 (bottom); Archaeopteryx Tours/ Shutterstock Images, 47; Carrie Olson/Shutterstock Images, 48; Sharif Uddin/500px/Getty Images, 49; imageBROKER/Rolf Nussbaumer/Newscom, 50 (left); Michael J Thompson/Shutterstock Images, 50 (right); M. Leonard Photography/Shutterstock Images, 51; Marga28/Dreamstime. com, 52; Ryan S Rubino/ Shutterstock Images, 54; Steve Byland/Shutterstock Images, 56, 71 (left); Sari ONeal/Shutterstock Images, 57; SweetyMommy/Getty Images, 59 (bottom); Gerald A. DeBoer/Shutterstock Images, 60, 68; Agami Photo Agency/ Alamy Stock Photo, 61; Bonnie Taylor Barry/Shutterstock Images, 62 (left), 85, 92 (right); Rabbitti/ Shutterstock Images, 62 (right); Photo Spirit/Shutterstock Images, 64; Rowdy Soetisna/Shutterstock Images, 66 (top); Makic Slobodan/ Shutterstock Images, 66 (bottom); Petr Simon/Shutterstock Images, 67 (bottom); Gregory Johnston/ Shutterstock Images, 70; Jaclyn Vernace/Shutterstock Images, 71 (right), 80 (right), 93 (left); Keneva Photography/Shutterstock Images, 72; John charles hansen/ Shutterstock Images, 73; John L. Absher/Shutterstock Images, 74; Fnach/Shutterstock Images, 76; Drakuliren/Shutterstock Images, 77 (right); karen burgess/ Shutterstock Images, 79 (left); Harold Stiver/Shutterstock Images, 79 (right); Wirestock Creators/Shutterstock Images, 80 (left); Sundry Photography/ Shutterstock Images, 82, 102; Kenneth D Carpenter/ Shutterstock Images, 83 (left); Henri_Lehtola/Shutterstock Images, 83 (right); Bob Pool/ Shutterstock Images, 86; Paul Tessier/Shutterstock Images, 87; Danita Delimont/ Shutterstock Images, 88; Feng Yu/Shutterstock Images, 90; Matthew Orselli/Shutterstock Images, 92 (left); Mark W. Holdren/ Shutterstock Images, 93 (right); Andrei Stepanov/Shutterstock Images, 97 (top); Danita Delimont/ Shutterstock Images, 97 (bottom); Kerry Hargrove/Shutterstock Images, 99; Agnieszka Bacal/ Shutterstock Images, 101; GE Condit/Shutterstock Images, 103; Wildred Marissen/Shutterstock Images, 104 (top); Birdiegal/ Shutterstock Images, 104 (bottom); Dominic Gentilcore PhD/ Shutterstock Images, 106

ABDOBOOKS.COM
Published by Abdo Reference, a division of ABDO, PO Box 398166, Minneapolis, Minnesota 55439.
Printed in China

102023
012024

Editor: Leah Kaminski
Series Designer: Colleen McLaren

Library of Congress Control Number: 2023939673
Publisher's Cataloging-in-Publication Data
Names: Golkar, Golriz, author.
Title: Songbirds / by Golriz Golkar
Description: Minneapolis, Minnesota : Abdo Reference, 2024 | Series: North American field guides | Includes online resources.
Identifiers: ISBN 9781098293109 (lib. bdg.) | ISBN 9798384911043 (ebook)
Subjects: LCSH: Songbirds--Juvenile literature. | Birds--Juvenile literature. | Birds--Behavior--Juvenile literature. | Birds--United States--Juvenile literature. | Encyclopedias and dictionaries--Juvenile literature.
Classification: DDC 598.2--dc23

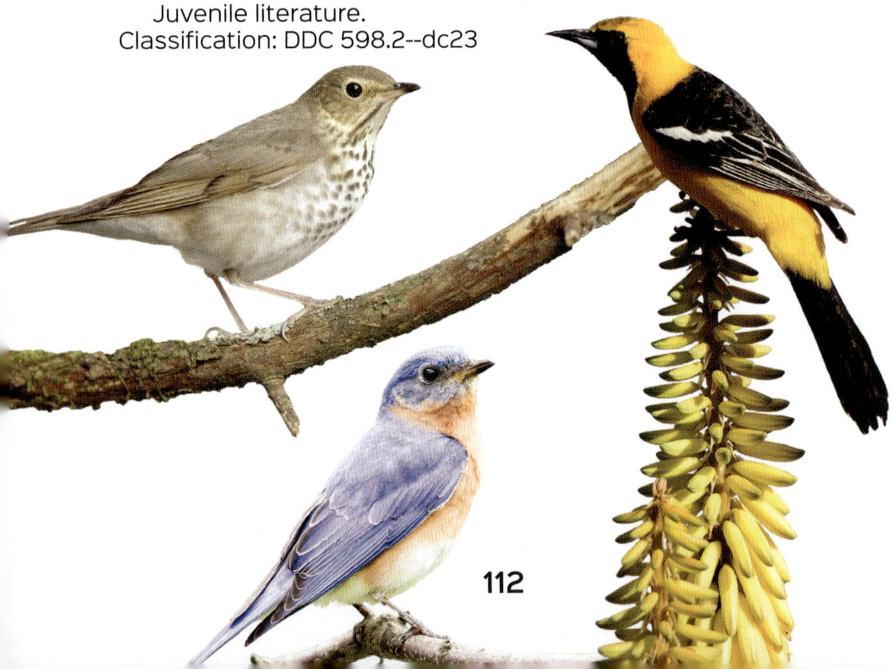